behavior – territoriality

GAZELLES AND THEIR RELATIVES

NOYES SERIES
IN
ANIMAL BEHAVIOR, ECOLOGY, CONSERVATION AND MANAGEMENT

A series of professional and reference books in ethology devoted to the better understanding of animal behavior, ecology, conservation, and management.

WOLVES OF THE WORLD: Perspectives of Behavior, Ecology, and Conservation.
Edited by *Fred H. Harrington* and *Paul C. Paquet*

IGUANAS OF THE WORLD: Their Behavior, Ecology, and Conservation.
Edited by *Gordon M. Burghardt* and *A. Stanley Rand*

HORSE BEHAVIOR: The Behavioral Traits and Adaptations of Domestic and Wild Horses, Including Ponies.
By *George H. Waring*

GAZELLES AND THEIR RELATIVES: A Study in Territorial Behavior
By *Fritz R. Walther, Elizabeth Cary Mungall,* and *Gerald A. Grau*

THE MANAGEMENT AND BIOLOGY OF AN EXTINCT SPECIES: *PÈRE DAVID'S DEER*
Edited by *Benjamin B. Beck* and *Christen Wemmer*

GAZELLES AND THEIR RELATIVES

A Study in Territorial Behavior

by

Fritz R. Walther
Elizabeth Cary Mungall
Gerald A. Grau

Texas A & M University
College Station, Texas

Library of Congress Catalog Card Number: 82-22245
ISBN: 0-8155-0928-6
Printed in the United States

Published in the United States of America by
Noyes Publications
Mill Road, Park Ridge, New Jersey 07656

10 9 8 7 6 5 4 3 2 1

Library of Congress Cataloging in Publication Data

Walther, Fritz R.
Gazelles and their relatives.

(Animal behavior, ecology, conservation, and management)
Bibliography: p.
Includes index.
1. Gazelles--Behavior. 2. Territoriality (Zoology) 3. Mammals--Behavior. I. Mungall, Elizabeth Cary. II. Grau, Gerald A. III. Title. IV. Series.
QL737.U53W326 1983 599.73'58 82-22245
ISBN 0-8155-0928-6

Preface

The numerous gazelle species and their relatives such as blackbuck, springbok, gerenuk, form the subfamily Antilopinae (p. 1) within the large family of horned ungulates (Bovidae). Their beauty and grace has attracted the attention of humans since ancient times. The Bible (for example, in the Song of Solomon) has no greater praise for the beauty of a woman than to compare it to that of a gazelle. Likewise, Indian artists have not tired of painting blackbuck since the times of the Moghul rulers. At least some of the species under discussion were known to science since the days of Aristotle. However, not very much has been known of their behavior, and generally, reliable information on the behavior of non-domesticated horned ungulates remained scarce up to the middle of this century. Most data were widely scattered in popular books by hunters and travelers. In these accounts observed facts, second-hand reports, and interpretive assumptions were often indistinguishably and uncritically mixed. The few attempts by scientists resulted either in isolated monographs of the behavior of a species, such as McHugh's (1958) work on bison *(Bison bison)*, or in monographs on one isolated behavior pattern, such as Schneider's (1931) studies on *Flehmen*. Only Antonius (1939) and, above all, Hediger (1941, 1942, 1949, 1951, 1954) pioneered a broader approach. Beginning in 1958, research intensified and became systematized on a comparative basis; however, the number of researchers remained small; there were only about ten in the whole world for another ten years. Only recently has the study of the behavior of wild bovids shown signs of the explosive development characteristic of behavioral sciences nowadays.

After these remarks, nobody will be surprised that there are still considerable gaps in our knowledge. For example, the subfamily of the duikers (Cephalophinae) comprises at least 15 species, but only one of them *(Cephalophus maxwelli)* has been investigated behaviorally to some extent (Aeschlimann 1963, Ralls 1969, 1974, 1975). By contrast, gazelles and their relatives (Antilopinae) are a comparatively well-investigated group. The following species have been studied behaviorally: Indian blackbuck in captivity (Hediger 1941, Backhaus 1958, Walther, 1958, 1959, 1968a, Benz 1973, Schmied 1973, Mungall 1979), and in the wild (Schaller 1967, Cary 1976b, Mungall 1978a,b, Prasad 1981), spingbok in the wild (Bigalke 1970, 1972, Mason 1976, Walther 1981), dibatag in captivity (Walther 1963a) and in the wild (Meester 1959), gerenuk in captivity (Backhaus 1958, Walther 1958, 1961, 1968a, Leuthold 1973) and in the wild (Ullrich 1963, Leuthold 1971, 1977, 1978a,b), Grant's gazelle in the wild (Walther 1965, 1968a, 1972a,b, 1977b, Estes 1967), Sömmering's gazelle in captivity (Walther 1964b), dama gazelle in captivity (Mungall 1980), mountain gazelle in the wild (Grau 1974, Grau and Walther 1976), dorcas gazelle in captivity (Walther 1966, 1968a), Thomson's gazelle in captivity (Walther 1958) and in the wild (Brooks 1961, Walther 1964a, 1967, 1968a, 1972b, 1973a, 1977a, 1978a,b,c, Estes 1967), Speke's gazelle in captivity (Walther 1958), and goitered gazelle in captivity (Walther 1963b, 1968a).

Dibatag, Speke's gazelle, Sömmering's gazelle and dama gazelle were observed under conditions which did not allow positive conclusions on territoriality; though there were some indications for it, at least in dibatag and Sömmering's gazelle. In the dorcas gazelle, territorial behavior is very likely according to the data available. In Indian blackbuck, springbok, gerenuk, Grant's gazelle, mountain gazelle, Thomson's gazelle, and goitered gazelle, there is no doubt. Thus, it is not too far-fetched to expect that the other species of this group may also be found to be territorial when they are eventually investigated.

In all the species studied so far, the following generalizations can be made: (a) only adult males become territorial, but not all of them, (b) usually territorial periods alternate with non-territorial periods during the life of the same individual, (c) the owners aggressively exclude other males from their territories, or at least dominate them within the territorial boundaries, and (d) usually the females only temporarily visit the males in their territories.

Marking with preorbital gland secretion and/or with urine and feces seems to play a role in the territoriality of some of the Antilopinae species. The territorial behavior of the males also influences the social organization and the spatial distribution of the other conspecifics. Above all, this territoriality is of greatest importance to reproductive success since, in all these species, reproduction is almost exclusively with the territorial males.

Thus, the occurrence and importance of territorial behavior in gazelles and their relatives have been fully recognized. On the other hand, the information published to date does not go much beyond the sketchy outline above. Little or no information is available on the relations of environmental conditions with territoriality, duration of territorial periods, territory size and structure, social situations leading to territorial establishment, phases in territoriality, conditions under which a male gives up his territorial status and abandons his territory, re-occupation of the same territory by the same male after a period of absence, the frequency and readiness with which other males take over an abandoned territory, the marking system, different forms of aggression used by a territorial male in encounters with males of different age (adults or subadults) or of different social status (territorial neighbors or non-territorial "bachelors"), and many other details. Considering the great importance of territorial behavior in the life of these animals, this kind of detailed knowledge is important with respect to not only theoretical concepts, but also practical problems such as wildlife conservation in national parks, management of exotic species on game ranches, and keeping and breeding such animals in zoological gardens.

Recently, researchers at the Department of Wildlife and Fisheries Sciences, Texas A & M University, have studied several Antilopinae species extensively enough to allow statements on just such details of territorial behavior as are listed above. Elizabeth Cary Mungall investigated territorial behavior of Indian blackbuck on several game ranches in Texas and later, in India. Gerald A. Grau studied mountain gazelle in Israel. Fritz R. Walther gathered data on Grant's and Thomson's gazelles in East Africa and on springbok in South West Africa. Together with the information on other species and of other researchers, data from these studies may give a view of Antilopinae territorial behavior as detailed and comprehensive as possible at present. The purpose of this publication is to make this information available, and thus, to contribute to the knowledge on behavior of several exotic game species important for wildlife conservation and management. Furthermore, we hope that this presentation may contribute to a better understanding of certain general problems of territoriality.

Numerous institutions and foundations have contributed to the investigations under discussion and the list of individual people who gave, often crucial, help to us is so long that it would fill pages. Therefore, we decided to name only those without whose assistance our research projects would have been impossible even to start.

We certainly have not forgotten those who gave us hospitality, transportation, and often invaluable practical advice, who searched for us when we had become lost or stuck in the field, who repaired our cars, who helped us to overcome the many nitty-gritty formalities involved in such research projects, who acted as translaters and interpreters when our foreign language vocabulary was at its end, etc. We also know how much we owe our teachers and the pioneers in our field of science who inspired us in our research and provided the scientific basis from which we could start. Last, but not least, we know how much we owe our friends and relatives, parents, wife, and children, who gave us their love and did not let us down even when they did not see us for years. We have not forgotten them and their devotion, but we ask for their forgiveness when we do not name them here. The list would have seemed endless.

F.R. Walther's studies on Thomson's and Grant's gazelles were financed mainly by the Fritz Thyssen Stiftung (Köln, Germany). Additional support was given by the Gertrud Rüegg Stiftung (Zürich, Switzerland), the Deutsche Forschungsgemeinschaft (Bad Godesberg, Germany), the Smithsonian Institution (Washington, USA), and the Caesar Kleberg Research Program in Wildlife Ecology (College Station, USA). The Zoologische Gesellschaft von 1858 zu Frankfurt a. M. (Germany) provided a four-wheel drive vehicle (Land Rover). F.R. Walther feels particularly obliged to Dr. H. Hediger (Zürich), Dr. B. Grzimek and Dr. R. Faust (Frankfurt a. M.), Dr. K. Lorenz and Dr. W. Wickler (Seewiesen), Dr. P. Glover (formerly Nairobi), Dr. T. Mcharo (formerly Seronera), and Dr. J.G. Teer (formerly College Station). The permit to work in Serengeti was given by the (former) directors of the Tanzania National Parks, Dr. J. Owen, S. Ole Saibull, and D. Bryceson. The study would have been impossible without the dedicated work of the game wardens and game scouts who made and kept the Serengeti National Park going.

E. Cary Mungall's investigation on blackbuck behavior in Texas was initially sponsored by a National Science Foundation Traineeship. The Tom Slick Graduate Research Fellowship Foundation funded one year of the field work, and the Caesar Kleberg Research Program in Wildlife Ecology carried all the rest of the project. E. Cary Mungall's blackbuck studies in India were made possible by the United States Fish and Wildlife Service through U.S. owned funds of the Special Foreign Currency Program. E. Cary Mungall wishes to extend a special word of thanks to Dr. J.G. Teer and to C.W. Ramsey (College Station) who coordinated the start of the project.

G.A. Grau's research on mountain gazelle was supported by Smithsonian Institution grants SFG-0-5181 and SFG-1-7066, and by the Caesar Kleberg Research Program in Wildlife Ecology. Additional support was given by the Ohio Cooperative Wildlife Research Unit, jointly supported by the Ohio Division of Wildlife, Ohio State University, U.S. Fish and Wildlife Service, and Wildlife Management Institute. He wishes to express special gratitude to Dr. H. Mendelssohn (Tel-Aviv), the personnel of the Tel-Aviv University and the members of Kibutz Nir Oz whose help and cooperation were essential to the study.

All three researchers owe the greatest gratitude to the Texas Agricultural Experiment Station and to the Department of Wildlife and Fisheries Sciences of Texas A & M University which took the protectorate role and enabled them to evaluate and write up their results. Also Dr. W.G. Klussmann (head of department) and Dr. F.S. Hendricks assigned an assistant, G.A. Dresner, to type the manuscript. Finally, a special word of thanks may go to Dr. W. Leuthold (Zürich) who kindly contributed Figures 26 and 34 for this book, and to K.S. Dharmakumarsinhji and S.E. Dougherty who contributed Figures 4c and 8d.

College Station, Texas
Lewisville, Texas
Washington, D.C.
June 1982

Fritz R. Walther
Elizabeth Cary Mungall
Gerald A. Grau

Contents

1

Introduction

ON CLASSIFICATION OF ANTILOPINAE

Gazelles and their relatives belong to the horned ungulates (family: Bovidae). The classification of these bovids has inherent problems. About one hundred years ago, they were divided into oxen, sheep, goats, and antelopes. The "antelopes" were defined by exclusion: all those horned ungulates which were not oxen, sheep, or goats were considered to be antelopes. Hence, a great number of heterogenous species were united under this term.

It is understandable that taxonomists later tried to improve this classification. Several different systems have been suggested, each with its own strengths and weaknesses, and none completely satisfactory. However, all have the same principle in that the former "antelopes" are separated into several subfamilies equivalent to the oxen (Bovinae), and the goats and sheep which have been united in one subfamily (Caprinae).

Within these new systems (e.g., Weber 1928), the gazelles and their relatives are a relatively well distinguished subfamily with the scientific name Antilopinae (= antelopes). Thus, it is somewhat misleading when hunters, travelers, etc., speak of antelopes and gazelles, as if they were different animals. According to the old classification systems, the gazelles belong to the antelopes, and in the modern classification systems, they even form the only subfamily for which this name has remained.

In 1945, Simpson included some small species in the Antilopinae subfamily, such as the dikdiks (genus *Madoqua*), klipspringer *(Oreotragus oreotragus)*, steenbok *(Raphicerus campestris)*, beira *(Dorcatragus megalotis)*, and others, which had previously been a subfamily of their own (Neotraginae). Thus, according to Simpson's classification, the subfamily Antilopinae consists of two tribes: the Antilopini (gazelles and their relatives, i.e., the former Antilopinae) and the Neotragini (dwarf antelopes, i.e., the former Neotraginae). It is not easy to judge the validity of this close unification. Simpson and his followers certainly had their reasons for it. On the other hand, there are obvious differences (as to size, color patterns, horn shapes, etc.) between the Antilopini and the Neotragini, and it seems to be more or less a matter of "taste" whether one puts more emphasis on the similarities or on the differences. With respect to behavior, one may safely say that some other bovids (e.g., the hippotragines, and here especially the genus *Oryx*) have about as many features in common with the gazelles as have the neotragines. For these and other reasons, we prefer the distinction of two subfamilies: Antilopinae and Neotraginae. Thus, when the term "Antilopinae" is used in the following, it refers exclusively to the gazelles and their relatives, but not to the smaller forms (the Neotraginae, or Neotragini).

In earlier classification systems, saiga *(Saiga tatarica)* and Tibetan antelope *(Pantholops hodgsoni)* belonged to the Antilopinae. Simpson (1945) united both of them with the Caprinae. Later, other authors considered them as a subfamily of their own (Saiginae). Haltenorth (1963) took the impala *(Aepyceros melampus)* out of the Antilopinae, and Ansell (1971) and Gentry (1971) established a separate subfamily (Aepycerotinae) for it. Considering behavioral aspects, the exclusion of impala from the Antilopinae appears to be very justifiable. Sokolov (1953) separated the dibatag *(Ammodorcas clarkei)* from the gazelles and united this species with the reedbucks (Reduncinae, or Reduncini in other classification systems). Apparently, a superficial resemblance of the dibatag's horns to those of the reedbucks gave rise to this change, but it certainly is an error.

Within the genus *Gazella,* the taxonomic positions of *"Gazella isabella," "Gazella saudiya," "Gazella arabica," "Gazella albonotata," "Gazella marica," "Gazella bennetti," "Gazella tilonura," "Gazella cuvieri,"* and *"Gazella rufina"* created particular problems. Investigations within the last 40 years (Schwartz 1937, Morrison-Scott 1939, Allen 1939, Ellermann and Morrison-Scott 1951, Boetticher 1953, Haltenorth 1963, Gentry 1964, Groves 1969, Lange 1972) lead to the view that these are not independent species, but subspecies of several different species. Although there are still a few controversies, most authors now agree that *"isabella"* and *"saudiya"* are subspecies of *Gazella dorcas, "tilonura"* and *"rufina"* are subspecies of *Gazella rufifrons, "cuvieri," "arabica,"* and *"bennetti"* are subspecies of *Gazella gazella, "albonotata"* is a subspecies of *Gazella thomsoni,* and *"marica"* is a subspecies of *Gazella subgutturosa.* In summary, the following species are relatively unquestioned as members of the Antilopinae subfamily (or the Antilopini tribe):

Tibetan gazelle, *Procapra picticaudata* Hodgson, 1846

Mongolian gazelle, *Procapra gutturosa* (Pallas, 1777)

Indian blackbuck antelope, *Antilope cervicapra* (Linné, 1758)

Springbok, *Antidorcas marsupialis* (Zimmermann, 1780)

Dibatag, *Ammodorcas clarkei* (Thomas, 1891)

Gerenuk, *Litocranius walleri* (Brooke, 1878)

Grant's gazelle, *Gazella granti* Brooke, 1872

Sömmering's gazelle, *Gazella soemmeringi* (Cretzschmar, 1826)

Dama gazelle, *Gazella dama* (Pallas, 1766)

Mountain gazelle, *Gazella gazella* (Pallas 1766)

Dorcas gazelle, *Gazella dorcas* (Linné, 1758)

Red-fronted gazelle, *Gazella rufifrons* (Gray, 1846)

Thomson's gazelle, *Gazella thomsoni* (Günther, 1884)

Speke's gazelle, *Gazella spekei* (Blyth, 1863)

Pelzeln's gazelle, *Gazella pelzelni* (Kohl, 1886)

Loder's gazelle, *Gazella leptoceros* (F. Cuvier, 1842)

Goitered gazelle, *Gazella subgutturosa* (Güldenstaedt, 1780)

Within the genus *Gazella,* there are three subgenera: The subgenus *Nanger* contains dama gazelle, Sömmering's gazelle, and Grant's gazelle. The subgenus *Trachelocele* has only the goitered gazelle. The subgenus *Gazella* includes all the other *Gazella* species.

DEFINITION OF TERRITORY

Territoriality includes several, difficult problems which have resulted in considerable confusion in the literature. Part of this confusion is due to the fact that it has not been possible to give a theoretically satisfactory, practically applicable, and generally accepted definition up to now, even though there have been many attempts by outstanding scientists. We are not bold enough to hope that we can settle all these problems; however, a short discussion is necessary to point them out and to explain how the terms "territory" and "territoriality" are used in this book.

One of the oldest definitions is Noble's (1939) statement that a territory is "any defended area." This definition is short and simple, and it clearly refers to the combination of a spatial aspect (the territory is an area, i.e., a specific place) with a behavioral aspect (defense, i.e., a form of aggression). For these reasons, Noble's definition may still be the best for introducing a freshman in behavioral sciences to the phenomena and problems of territoriality. On the other hand, for many reasons, this definition is not good enough for an advanced student of animal behavior, and, above all, not for a researcher in the field.

Its greatest weakness is that it does not allow a distinction between territoriality and non-territorial defense of an area. The latter definitely exists. For example, in Thomson's gazelle, a half-grown fawn may climb a termite mound and defend it against its playmates. Young ibex *(Capra ibex)* may show the same "king-of-the-castle" game over a rock (Walther 1960b). Or, particularly in captivity, an animal may defend the area around a preferred food or around a feeder. Even females of many species—including ibex, markhor goat *(Capra falconeri),* and others—show this behavior. In all these cases, and in a lot more, animals clearly defend an area, and thus, Noble's definition of territoriality is fully applicable to them. Nevertheless, the individuals involved either belong to age or sex classes (juveniles, females) or to species (e.g., ibex and markhor goat) which are not territorial. Moreover, play and food competition are not motivations for territoriality in ungulates. Therefore, the first objection to Noble's and similar definitions is inaccuracy in the definition itself resulting in difficulties in distinguishing the phenomena under discussion from others. Even if there were no other objections, this would be enough since it is the purpose of a definition to separate clearly certain phenomena from others. However, there are more objections.

In the fighting techniques as well as in the aggressive displays of many animals, including Antilopinae species, it is often possible to distinguish between offensive and defensive behavior, besides other techniques which may be used both ways. The phrase "a territory is a defended area" implies that a territorial animal shows defensive behavior, but the opposite is true. The territorial owner behaves offensively and dominantly on his own ground (Schenkel 1966). Thus, one can speak of territorial "defense" only in that paradoxical sense in which a soldier or an athlete may say that a good offensive is the best kind of "defense."

Moreover, there are cases, occasionally also in Antilopinae, in which a territorial male temporarily tolerates other males in his territory—as long as they behave subordinately. In such cases, we do not have any territorial defense, but only a territorial dominance. This marks our next objection that the concept of defense in the definition of territory is not broad enough. For example, in Grant's gazelle, with their relatively large territories, there are cases in which one male occupies the only suitable place in a given area (e.g., a clearing in a woodland). In such a case, there are no territorial neighbors present, and it may take weeks or months until a bachelor group happens to pass through the area. Thus, no territorial defense can be observed for long periods, and thus, any definition which is based exclusively on defense, or any other form of aggression, is inapplicable. This application problem certainly is the greatest weakness of such a definition of territoriality in its practical use.

On the other hand, it definitely is possible to determine the territorial status of an animal in such a case because, at least in ungulates, there are other behavior patterns, which are as indicative of territoriality as is aggression. These patterns include efforts to herd back females when they intend to leave the territory, higher thresholds for flight (from predators, cars, and people) inside the territory, etc. When such features are included in a definition, its practical application becomes considerably enhanced. Moreover, the existence of such features strongly indicates that territorial defense (or aggression, or dominance) is only one behavioral "symptom" among others of the animal's territorial status. Therefore, a definition which refers exclusively to defense is totally one-sided. Just as a good medicament should not only fight the symptoms of a disease but cure the process which causes the symptoms, in an analogous way, a good definition should name that factor which brings the behavioral symptoms about, not merely list the symptoms, or even only one of them.

Apparently, this basic theoretical problem has not been recognized by the majority of authors who later tried to modify the definition of a territory or of territoriality. Therefore, these trials have not brought truly significant improvement. It would take too long to give a complete review of all these definitions here. Suffice it to say that two principal ways have been tried.

First, some authors (e.g., Emlen 1957, Schenkel 1966, Leuthold 1977) tried to modify and broaden the concept of "territorial defense" by replacing these words with terms such as "territorial dominance," "intolerance," "negative reaction to other individuals," "exclusion of individuals of comparable social status by active repulsion," "preventing certain other individuals from engaging in certain activities," etc. As is clear from the preceding discussion of Noble's definition, these attempts are well justified, in so far as they provide a certain improvement. However, the outlined limitation in the practical use of the definition as well as the criticism of one-sidedness, of inaccuracy in the definition itself and of a somewhat superficial treatment of the problem (i.e., referring to symptoms instead of to a basic factor) remain fully valid.

The other type of modification suggested by the same or other authors (e.g., Nice 1941, Burt 1943, Etkin 1964) was to change the term "area." Phrases such as "positive reaction to a particular place," "fixation in space," "spatially fixed area," "part of an animal's home range," have been used. It is hard to see any significant progress with respect to the definition of territory and territoriality in these new wordings. Certainly, they are more elaborate and sometimes also more precise than the old term "area," but, on the whole, they are not much more than rephrasings. Moreover, an overly strong and somewhat one-sided emphasis on the spatial aspects of territoriality includes the danger that the distinction between territory and home range may be lost. Consequently, Etkin (1964) considers this distinction unnecessary. At least in animals such as Antilopinae, however, some of the males may stay in territories, but the non-territorial males and the females may display home range behavior in the same area and at the same time. Thus, the distinction between territories and home ranges, as difficult as it may be with respect to definition, is absolutely necessary. Perhaps, this distinction may even include the key to a better understanding and a better definition of the terms "territory" and "territoriality."

The deciding point seems to be that only a territory has a boundary (Walther 1972b). Unfortunately, there are two sources of misunderstanding to this formulation. Some people (e.g., Owen-Smith 1975) have the opinion that the term "boundary" would imply a sharp line determined to the precise meter, or even centimeter. This is not necessarily so. The term "boundary" can also be applied to a "zone," i.e., a broader "belt." More serious are terminology problems which arise from the broad sense in which the word "boundary" can be used in common language. Here, it often is synonymous with "end," "limit," etc., and thus, one may speak of the boundary of a forest, the boundary of a grassland, the boundary of the distribution of a species, etc. With respect to territories, however, the term "boundary" has to be used in a more special sense comparable to that in which we speak of the (political) boundary of a nation. This situation is different from a mere ending of a given area in that it is not due to environmental factors. Instead, it is subjectively established by the people, and it exists primarily in their imagination. Provided that such a political boundary is not artificially marked, there is nothing objectively present in the landscape from which its course or even its existence could be recognized. Thus, a stranger would pass it without notice, if not informed of its location. Secondly, such a boundary is not merely a result of habit, as, for example, when a person customarily restricts his excursions to a limited area. Although this man may not extend his walks beyond certain places, there is nothing which would prevent him from doing so. If he once does, he may perhaps experience feelings of insecurity due to entering unfamiliar terrain or to meeting strangers. However, even this is not necessarily so, and it would not lead to predictable, basic changes in his behavior as may occur in crossing a political boundary, e.g., when entering a country with different regulations and lifestyle.

In short, we wish to restrict the term "boundary" to cases in which (a) this boundary is subjectively established by the inhabitant(s), and (b) crossing this boundary results in significant and very definite behavioral changes of the inhabitant, which means that the behavior of the same individual is different in a number of aspects inside and outside his territory. This concept of "boundary" appears to be adequate to distinguish a territory from a home range and to approach a definition of territory which does not only list one or several behavioral peculiarities of the territorial status, but which tries to understand them as symptoms originating from one mutual basis.

Thus, with special reference to the situation in Antilopinae (and some other ungulates), we come to the following definition (Walther 1972b): A territory is a place in which an animal lives for a variable period of time and around which that animal has established a subjective boundary. This boundary can take the form of either a line or a zone. Only the animal which has established this boundary (i.e., primarily the owner of the territory and sometimes the immediate territorial neighbor) is aware of its existence. The territorial status of the owner as well as the existence of the territorial boundaries are indicated by, at minimum, the following behavioral symptoms: (a) intolerance of, or at least dominance over, conspecifics of the same sex within the boundaries; this intolerance being lost or strongly diminished outside the territory, (b) higher thresholds for flight (from predators, men, cars, etc.) inside the territory, (c) efforts to herd back conspecifics of the other sex when they attempt to leave the territory, (d) sudden halting of the territorial animal when arriving at the boundary (for example when herding females or pursuing trespassing males), (e) occasional occurrence of conflict behavior when the owner temporarily leaves his territory, and (f) establishment of a marking system related to the structure of the territory (marking the center and/or boundaries) in species which mark by olfactorial means such as feces, urine, or secretions of skin glands.

A home range is an area in which an animal, or, as almost always in Antilopinae, a group of animals stays long enough to establish a subjective space-time system, as it may also be found in a territorial animal, but a home range does not have subjective boundaries. Thus, the behavioral symptoms of territoriality listed above are lacking—except, perhaps, for a somewhat lower threshold for flight outside the home range. As a further consequence, the home ranges of neighboring herds may overlap, and when the herds are open societies, as in the Antilopinae, individuals may readily switch from one herd in one home range to another herd in another home range.

In some cases, it is practical to define a third type of subjective space which could possibly be termed an "action area." This space includes all seasonal territories and/or home ranges as well as the migration routes which connect them. In short, sometimes a special term (such as "action area") is needed to characterize the entire space on which an animal sets foot during its individual life. The same may also be applied to defined groups.

GENERAL METHODS

The basic method in our studies was to travel to an area well populated by the subject species and to remain there for several hours or days recording what good fortune allowed us to observe. This approach is not quite as opportunistic as it may appear, since it is based on the well-founded concept of spontaneity in animal behavior. "A healthy animal is up and doing," as McDougall (1923) phrased it. Thus, if sufficient time is spent with the animals, one can be sure of seeing behavior of interest to that specific study.

In these investigations, rather different methods have been used. Naturally, the techniques in mapping a territory are different from those for individual recognition, and both are different from those in determining group composition or in investigating daily activity rhythms of the animals, etc. It would be inadvisable to present all these methods and techniques in a long discussion of its own. In part, they will be described in connection with the corresponding data and results. In part, we are in the fortunate position of being able to unburden this text from merely methodological discussions by referring to previous publications which deal with them in detail (e.g., Walther 1969, 1972a, 1973a, 1977a, 1978a,b,c, Grau and Walther 1976). Much the same is true for evaluation techniques which ranged from frame-by-frame analyses of movie pictures to statistical evaluations of quantitative data.

We should mention that the three authors have not always used the same recording or evaluation principles such as categories of behavior patterns or of types of encounters. To a certain extent, we have tried to standardize them in this text, but some have been kept unchanged. In some cases, such differences reflect species-specific peculiarities, and sometimes variations in evaluation complement each other to provide a better general view.

Although some observations were made from fixed blinds, platforms, or on foot, the vehicle (Land Rover in Africa and in Israel, pick-up truck in Texas) was used as a "movable blind" in the majority of cases, except in India. In some of the areas, such as the East African national parks, the game animals were generally more habituated to the sight of cars than to pedestrians, and their flight distance from cars was considerably smaller than that from people. Apparently, they could not visually discriminate man from car, i.e., car and man visually formed one unit as long as the driver remained inside his vehicle. The greater shyness toward humans became evident at the moment they recognized the man as a separate figure, as when he left the car and stood beside it (Walther 1969). Furthermore, the use of the car had the great advantage that the observer could have all his equipment along and readily at hand. Also, he could follow the animals when they moved. Such movements of the car, however, were restricted to the minimum since the animals were more confident of a stationary vehicle than of a moving one. In an open plains situation, parking the car near an environmental object such as a single tree, shrub, rock, or termite mound, often made a difference. When the car was the only striking object in the surroundings, the animals kept a distance of several hundred yards for hours or days, whereas they "forgot" about the presence of the vehicle after a relatively short time if it was positioned in the immediate vicinity of a striking natural object. Flight releasing or facilitating factors, such as sudden appearance, fast speed, direct course toward the animals, and so on (Walther 1969) had to be avoided when approaching the animals. In short, when one arrived at a promising area, it was best to park the car near a natural object, to remain there throughout the observation time, and not to leave the car.

Good binoculars were very essential for our observations. The authors used 7 x 50 and 10 x 40 binoculars. For certain purposes, such as individual identification, a 40 x 60 telescope was preferred over binoculars. Usually, the observations were recorded with a tape recorder and later transferred to a notebook. Rough sketches drawn in the field were often extremely helpful. For documentation and for later analyses, still (35 mm) and movie (16 mm) pictures were taken whenever possible. All three researchers used telephoto lenses of f=400 mm and f=640 mm for their cameras.

SEX AND AGE CLASSES

Sex recognition in the field is fairly simple when dealing with animals older than the juvenile stage, i.e., older than six to nine months. Indian blackbuck females typically do not have horns, although a few horned females are known as abnormalities. In the majority of the gazelle species, females have horns, but they are much shorter and thinner than in the males. Even the small horns of young males are clearly sturdier than those of females. Furthermore, the thinner neck of females, their smaller size and lighter body, and in blackbuck adults often their color, may be used for sex recognition.

Under field conditions, sex recognition in juveniles was only possible when the animal was seen urinating. The urination posture is different in males and females, and this sex-specific difference occurs at a relatively early age (latest between two and three months). Occasionally, sex recognition was possible even in younger animals when one observed the mother licking her fawn. She reaches between a male fawn's hindlegs, under its belly, in order to lick the penis for stimulating urination (Walther 1966, 1973b), while her only cause to take a similar stance with a female fawn would be to lick the end of the umbilical cord during the first hours after birth.

Age class recognition for behavioral studies can be limited to relatively few categories which represent important phases in the animal's life. We used the following age classes: neonate, fawn age, half-grown, adolescent, subadult, young-adult, fully adult, and very old (Figures 1-4).

The neonate is completely dependent on its mother for nutrition. Also, urination and defecation are, at least, considerably facilitated by the mother's licking the neonate's anal and genital regions. The neonates of Antilopinae "lie out" (Walther 1965, 1966, 1968a) most of the time, and join their mothers only temporarily for nursing and cleaning. Besides their small size, in blackbuck as well as in Thomson's gazelle, neonates can be recognized by their darker color which gives way to the lighter color of an adult animal (in blackbuck, to a more creamy color than that of an adult female for about one month) at the latest by the end of the second week. Thus, the neonate category ranges from birth to an age of about two weeks.

Figure 1: Sex and age classes in Thomson's gazelle (a) Neonate fawn, being licked by mother (adult female). (b) Fawn with mother (adult female). (c) Half-grown fawn with female (possibly young-adult). (d) Adolescent female (right) and adult female with a deformed and a broken horn (presumably the mother). (e) Subadult female. (Compare horn length and body depth to adult female in Figure 1b.) (f) Adolescent male. (g) Subadult males. (h) Adult male. (Photos: F.R. Walther—Serengeti National Park, Tanzania.)

In the fawn age, the young are still small, although somewhat bigger than neonates. While standing beside its mother, the fawn's dorsal surface is below the mother's belly or level with it. Thomson's gazelle fawns often have a little white mane on the neck near the chest. Horns are not present or, at least, not visible under field conditions. Fawns still lie out, and are nursed and cleaned by their mothers. Thus, their behavior is largely the same as that of neonates; however, certain changes gradually occur. The length of the lying-out periods declines, and the fawn spends more and more time on its feet gambolling around its mother after nursing and cleaning. Occasionally, it may interact with conspecifics of the same age or somewhat older. Although the fawn still relies predominantly on the mother's milk, it begins to take solid food. The fawn stage ranges from two weeks to about two months.

Figure 2: Age classes in male Grant's gazelle (a) Young-adult male. (b) Fully adult male. (c) Very old male. (The bucks in Figure 2a,b, and c have horns of the Robert's type.) (d) Adolescent male. (e) Subadult male. (f) Females for comparison. On the left: adolescent doe, almost to the age of a subadult. On the right: adult doe, presumably the mother of this adolescent doe. (Photos: F.R. Walther—Serengeti National Park, Tanzania.)

Figure 3: Sex and age classes in mountain gazelle (a) Adult male. (b) Subadult male. (c) Adult female. (Photos: G.A. Grau—Northern Negev and Research Zoo, Tel-Aviv University, Israel.)

When standing beside an adult female, the dorsal surface of a half-grown fawn is approximately at the level of half of the adult's body. At the end of the half-grown stage, the tips of the horns usually are visible under field conditions; however, they do not yet allow a distinction of the sexes in species where both males and females have horns. Seen at close range, the forehead of a half-grown male appears to be somewhat broader than a female's of the same age. According to Robinette's and Archer's (1971) data on horn growth in Thomson's gazelle, this half-grown fawn class seems to correspond to an age of three to four, or perhaps even five to six months. The age range for this class in blackbuck is one to four months in females and one to five months in males (Mungall 1978a). The lying-out period is over. The young may still rest more frequently and for longer periods than the adults, but they now participate in adult activities and move around with their mothers. Thus, half-grown fawns are in female herds or mixed herds. The young of this age have converted to solid food but may still try to suckle, with or without success. Elimination is completely independent of the mother's licking.

Adolescents are still smaller than adult females and have a proportionally shorter snout. The horns of adolescent males are shorter than their ears or, at maximum, equal in length. In blackbuck, they usually curve forward or outward but lack the conspicuous twisting of older males. Although the horns of adult gazelle females are about the same length or even longer, the horns of adolescent males are much sturdier than those of females. At the beginning of this age class, they are about 3 cm tall and look like little cones. Later, they also show the typical rings of male horns. In Thomson's gazelle, Robinette and Archer (1971), Hvidberg-Hansen and de Vos (1971) as well as Walther (1973b) agree that this class corresponds to an age of five to eight months. Adolescent blackbuck males are six months to one year old (Mungall 1978a).

Figure 4: Sex and age classes in blackbuck (a) Adult female with her fawn which still has the color of a neonate. (b) Left to right: subadult, adolescent, and adult male. (Photos: E.C. Mungall—Y.O. Ranch, Texas.) (c) Young-adult and fully adult males. (Note coat color variation among adult males.) (Photo: Dharmakumarsinhji—Velavadar Blackbuck National Park, India.)

The horns of adolescent females may or may not be visible under field conditions in Thomson's gazelle and mountain gazelle. In Grant's gazelle, they usually are visible. They are much thinner than in adolescent males and reach, at maximum, three quarters, but in most cases only one third, of the length of the ears. In Thomson's gazelle, the little white mane near the chest can still be present. There seem to be great individual variations, and, exceptionally, it may still be present in adult females. Thus, in many cases, only the smaller size and the somewhat different body proportions are the criteria for distinguishing an adolescent female from an adult.

Adolescents of both sexes are weaned but occasionally may still suckle or, at least, try to do so. Adolescent females are frequently seen near adult females, presumably their mothers. Occasionally, adult males may sexually approach or herd adolescent females, whereas they usually pay little or no attention to fawns and half-grown fawns. In mountain gazelle, Mendelssohn (1974) found about 73% of the one-year-old females to be pregnant, i.e., they might have been bred while adolescents.

Territorial males may sometimes take offense at the presence of adolescent males in their territories and chase them. Circumstances determine whether this results in permanent separation of the youngster from his mother. If separated, the adolescent male will join a bachelor herd which also offers him more opportunity for playful sparring than a female herd. Consequently, adolescent males are found in female herds as well as in all-male groups and, of course, in mixed herds. Adolescent males show most of the behavior patterns of adult males; however, some of them (e.g., certain dominance and courtship displays) are infrequent or less pronounced (e.g., the urination and defecation sequence). Adolescents are not territorial. Although they occasionally display interest in females, they have not been observed copulating. Adult females may defend themselves successfully against sexual approaches of adolescent males.

With the exception of the horns, the appearance of a subadult male largely corresponds to that of an adult female. The neck is still clearly thinner than in an adult male, but often slightly thicker than that of a female. The horns are longer than the ears; however, they have not yet reached their complete size and final shape. In all the gazelle species under discussion, the horns of subadult males show a "C" shape, i.e., the lower curve which gives an "S" shape to the horns of an adult buck, has not yet developed. In blackbuck, the coats of many subadult males start to darken, especially on the face, and they have horns of almost two to barely three twists, whereas adults have horns of three twists or more. In this case, "twist" means a full curve of the horns to the outside. In Thomson's gazelle, the subadult males apparently correspond to Robinette's and Archer's (1971) age classes "V" and "VI" for which the authors give an age of eight to twelve months. However, among Walther's (1973b) collection of skulls from predator kills, there was a skull of a subadult male which seemed to correspond to an age of 14 to 15 months. Also, a 15-month-old mountain gazelle in the research zoo of the Tel-Aviv University was in this age class, and in Grant's gazelle, subadult males were generally estimated to be between one to two years (Walther 1972a). In blackbuck, the trend for faster development in females than in males continues through the subadult stage. Subadult females are six months to one year old whereas a subadult male is one to three years old.

Suckling attempts were never observed in subadult animals. Subadult males are usually not found in female herds. They are in all-male groups or in mixed herds. They do not try to become territorial. They are dominant over adult females and may show full mating behavior; however, copulations are extremely exceptional (one observation in blackbuck). The subadult males probably have reached sexual maturity but cannot mate successfully under normal conditions because they are still unable to establish territories.

Subadult females are difficult to distinguish from adult females. Their horns may have half the length to about equal length of the ears. Subadult females are a little smaller, the body and the neck appear to be more slender, and the nose region shorter relative to the forehead than in adult females. Recognition of subadult females is relatively easy in Grant's gazelle. In Thomson's gazelle and blackbuck, the subadult age of females can only be recognized in observations at close range and even then only when fully adult females are around for an immediate comparison. In mountain gazelle, it is usually impossible to distinguish subadult from adult females under field conditions. Subadult females are courted by males, and copulations have been observed.

Recognition of young-adult females, i.e., females having reached full maturity only recently, was impossible in all the species under field conditions. Recognition of young-adult males was not easy either, but not as completely out of the question as with females. A young-adult male has reached the full male size. The neck is clearly thicker than that of a female, but it is still somewhat thinner than that of a fully adult male. Also, the body is not as heavy as in a fully adult buck. In Thomson's gazelle, the horns are about 20 cm long, and they show the beginning of the lower curvature which, at maximum, approximates the length of the upper curvature (of the "S"), but usually it is shorter than the upper curvature. In older males of this species, it is the opposite. In Grant's gazelle of the Robert's type *(Gazella granti robertsi),* which is common in the Serengeti area, the horns of a young-adult buck do not yet show the complete curve to the rear and, above all, they do not yet curve sideward as is typical of adult *robertsi.* For these reasons, the horns of a young-adult Grant's buck appear to be very steep, especially when seen in profile. In Thomson's gazelle, the examination of two skulls of males of this age class lead to the estimation of more than 15 months, but not yet two full years (Walther 1973b). In Grant's gazelle, the young-adult males were estimated to be between two and three years (Walther 1972a). In blackbuck, young-adult males tend to have horns of just three twists, but there is too much individual variation for this to be a reliable criterion. Young-adult males are of some special interest in that males of this age may first try to establish territories, although often without success.

In Thomson's gazelle and probably also mountain gazelle, full adulthood seems to be reached with the end of the second year, in Grant's gazelle during the third year and in blackbuck by the third year in females and by the fourth year in males. A striking field characteristic of adult males in all species under discussion is the thick neck. In Indian blackbuck, the orange-tan body color of juveniles, females, and adolescent males may change to brown or black. However, it must be mentioned that some adult blackbuck males remain orange-tan except for the black on the face. In part, these differences in coloration seem to be linked with social status, e.g., tan adult males are more likely to be non-territorial males. In part, the colors also vary with season, i.e., some of the adult blackbuck males are only dark in winter. In Thomson's and mountain gazelles, the horns of adult males have the full "S" shape. Their length is 25 to 43 cm in Thomson's gazelle, and 25 to 37 cm in mountain gazelle (Haltenorth 1963). In Grant's gazelle, the horns also show a curvature similar to an "S," and in the Robert's type, their upper two thirds turn somewhat sideward-outward with the tips bending somewhat down (the last characteristic is already valid for young-adult males). Horn length ranges from 50 to 80 cm in adult Grant's bucks (Haltenorth 1963). Indian blackbuck horns show three to five—very occasionally six—twists, and they are usually 33 to 74 cm in length (Mungall 1978a).

Fully adult Antilopinae males may become territorial and service females. Alternatively, they may stay in all-male groups, often making up the majority of the members in these groups, or, especially during migratory periods, they may run in mixed herds. As far as there is object marking in the species, they frequently mark with their preorbital glands, and they urinate and defecate in striking postures and in sequence. Scraping the ground with a foreleg may or may not precede urination and defecation; however, it was never observed in males of a young age. Adult bucks show the full display repertoire of their species. There are no qualitative differences in the behavior of territorial and non-territorial adult males on the level of fixed action patterns. Quantitative differences do exist as will be discussed later in detail. Thus, the territorial or the non-territorial status of an adult male can only be recognized by applying the complex territoriality criteria (p. 6).

The distinction between adult males and females is very easy due to the much thinner neck and the much shorter horns of the does. In blackbuck, as well as in gerenuk, dibatag, and frequently also in goitered gazelle, females have no horns at all, and blackbuck females also differ in color from many adult males. In the gazelle species, the horns of adult females have approximately the thickness of a pencil, a carpenter's pencil in Grant's gazelle, and they neither show the typical curves of male horns, nor their pronounced rings. The length of female horns is between 10 and 20 cm in mountain gazelle and in Thomson's gazelle it ranges from 8 to 15 cm (Haltenorth 1963). The females' horns in both these species are often broken, bent, or otherwise deformed. As compared to the other gazelle species, Grant's gazelle females have the longest horns, ranging from 30 to 43 cm, and deformations of their horns are rare.

Not too much is known about aging and the lifespan of Antilopinae species. In captivity, blackbuck quite frequently live 11 to 12 years (Dharmakumarsinhji 1967, Schmied 1973) and the record is near 16 years (Crandall 1964). Likewise, two male gerenuk of known age lived 11 and 12 years in the Zoological Garden of Frankfurt a. M. (Faust pers. comm.). Generally, by about 10 years, gazelles and their relatives seem to reach an age which may be termed "very old" (Walther 1973b). The belly hangs deep in such animals, the ribs are often visible, and sometimes the lower lip appears somewhat protruded. In Thomson's gazelle, the normally brown forehead (often with a small white patch in its center) may become completely white, framed only by small stripes of darker color.

Old blackbuck whose teeth are worn to stubs, are grizzled on the neck and even more on the head. In very old Grant's bucks of the Robert's type, the lower, vertical part and the upper, back and sideward curved part of the horns are of about equal length, forming an angle of approximately 90°. In Thomson's and Grant's gazelle, a few males who showed all the described features of a very old age, were observed to be territorial. It so happened that none of them were seen copulating. However, a captive 12 year old blackbuck male, who had lost his dominant status about five months before, successfully bred a female a few weeks before he was destroyed because of his weakening condition (Schmied 1973).

INDIVIDUAL IDENTIFICATION

The study of territoriality always includes the problem of individual identification because of the importance of determining the length of the individual's stay in a territory, the re-occupation of a territory by the same male, etc. Artificial marking, of course, is a possibility; however, it necessarily includes capture or immobilization of the animal. During such procedures, the risk that the animal may be killed or crippled can never be excluded, particularly with species as small and delicate as gazelles and their relatives. In addition, the marks, e.g., collars, may cause problems for the animals and alter their behavior. Furthermore, such actions are always expensive and time consuming. For these as well as other reasons (Walther 1973b), a field researcher should use artificial marking only in those cases in which he cannot obtain satisfactory data in any other way. We did not artificially mark our animals.

With respect to individual identification, the study of territorial behavior offers two considerable advantages *a priori:* the investigator studies only a limited number of individuals and only at a definite and limited place. Under these circumstances, certain natural and relatively simple features allow a reasonably sure identification in male Antilopinae.

One feature is provided by the horns which often vary in length and shape within the normal range. For example, they can be relatively long, of comparatively weak curvature, and almost parallel, or in another individual, they may be relatively short, of comparatively weak curvature, and diverging in a wide V-shape; in a third male, they may also be V-shaped, but they may be long and may show a very pronounced curvature, etc. In addition, the horn tips may be sharp or blunt. Quite frequently, there are also small irregularities such as the right horn bent somewhat more backward than the left one (or vice versa), or the one horn bent somewhat more to the outside than the other, not to mention occasional abnormalities such as broken or deformed horns. Particularly in Indian blackbuck and in Grant's gazelle, such individual horn variations within the normal range are common and often very pronounced.

In addition, in mountain and Thomson's gazelles, facial marks can be used for individual identification. For example, in Thomson's gazelle (Figure 5), there are great individual variations in size and shape of the black patch on the nose (triangle, spindle, double-axe, etc.) and of the white patch within the otherwise dark-brown forehead (pear-shape, fork, circle, etc.). Although size and shape of this white forehead patch may change with age, this does not matter in the observation of an individual during a territorial period of a few weeks or months. Since the black nose patch and the white forehead patch vary independently of each other, they provide an almost endless list of individual combinations (Walther 1973b). Furthermore, there sometimes are variations in size and shape of the dark lateral eye-stripe such as long, straight and thin, or, in another individual, short, thick and curved, or, in a third animal, long, thin and curved (Walther 1973b).

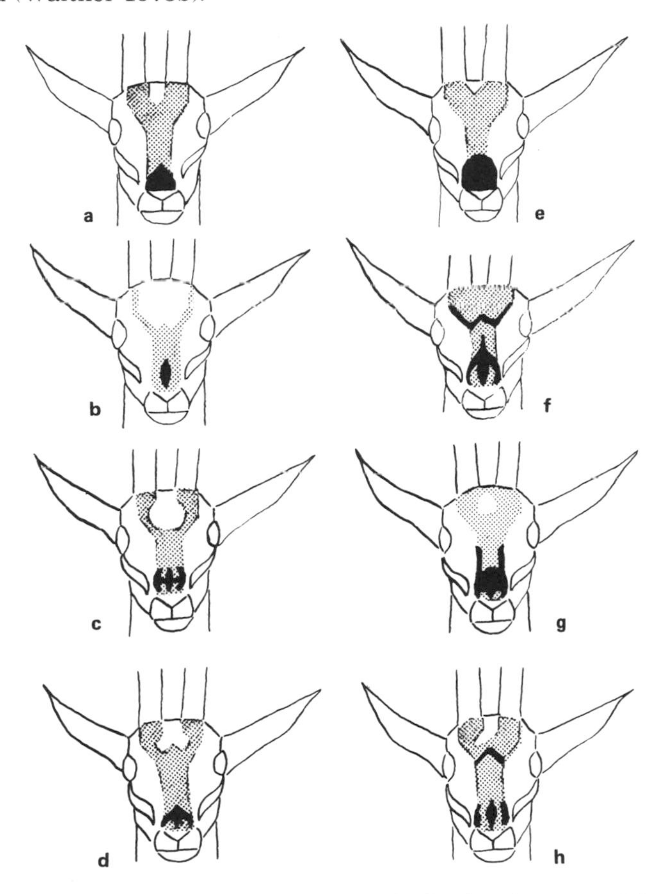

Figure 5: Some examples of variations in forehead and nose markings of Thomson's gazelle.

White Patch on the Forehead	Black Patch on the Nose	Additional Black Markings
a. Stripe	Triangle	—
b. White front	Spindle	—
c. Pear-shape	Double-axe	—
d. Fork	Tripod	—
e. Triangle	Broad, shapeless	—
f. Lacking	Big spindle with side appendices	"W" shape
g. Circle	Broad tripod	Paired vertical stripes
h. One-sided fork	Interrupted double-axe	Inverted "V" shape

In addition to horn size and shape and to the differences in facial markings, there are smaller but often well-recognizable variations in body color (darker—lighter), in size, and in general body appearance (heavy—slender). Among blackbuck males, general body color is difficult to use unless the same individuals are re-examined at least once a week because of the continual changes in coat color affected by seasons, rank changes, etc. However, there are sometimes differences in the angle of the dark-white border along the side. Also, variations in the tendency toward a white, partial chevron on the face from the eye rings sometimes can be seen by telescope. Occasionally, of course, there are also scars and minor physical abnormalities in all the species, e.g., an animal which has lost half of its tail or its ear.

Moreover, behavioral peculiarities sometimes characterize individual animals. For example, one of the territorial Thomson's gazelles in Serengeti could regularly be approached within 5 m. All the others fled at the latest when approached within 20 to 30 m. Another of these males apparently had some trouble with his ears (possibly parasites) and scratched them and shook his head considerably more often than gazelles normally do. Also, one of the territorial blackbucks in a Texas pasture frequently scratched himself with a hindleg bent perpendicularly at the hock, a method rarely seen in other individuals.

In short, there are a great number of natural physical and behavioral features which can be used for the identification of individuals, at least, as long as they remain in circumscribed and relatively small areas as territorial Antilopinae males do. Frequently, one of these features alone is not good enough for a sure identification, but when one uses several of them in combination, they work quite well. Finally, it must not be forgotten that each of the individual animals under discussion was observed, at minimum, for about 40 hours, but most of them were observed for several hundred hours, and "observation" means in this case that the researcher literally did not take his eyes from the animal. During such long and intensive observations of one individual, the investigator becomes as familiar with its appearance and can distinguish it as reliably from others as his own dog, cat, horse, or human friends who usually are not observed half as long and intently and are not marked either.

2

Animal Populations and Habitats; Study Areas

INDIAN BLACKBUCK

Indian blackbuck were once dispersed locally throughout India wherever conditions were favorable (Jerdon 1874), and their number may have approximated four million (Groves 1972). Hunting and habitat destruction have brought them down to about 10,000 in India. In their native country, blackbuck are essentially a plains species. They inhabit scrub and grassland, but may also penetrate into the more open parts of predominantly deciduous forests formed by stunted teak, and various other small trees (Prater 1971). Blackbuck thrive best in areas where the vegetation is not too dense, the climate not too moist, the temperature does not fall too low and the topography is not too rugged. Therefore, the blackbuck's maximum known distribution in India covers the northern, central, and southern plains and open woodlands but is limited by the thick jungles of the west coast, the wet portions of Bengal and Assam and the cold, steep slopes of the Himalayas.

The Indian blackbuck antelope is one of several exotic ungulates that have been widely introduced onto Texas ranchlands since the 1930s and have established vigorous populations there. The original stock came from zoo surpluses. In a 1974 census, 7,339 blackbuck were counted on Texas ranches with over 80% of them living on the Edwards Plateau (Harmel 1975). The major industry on the Edwards Plateau is cattle, sheep, and goat ranching. In addition, hunting native and exotic game has grown into an economically important industry that supplements ranch income. The conspicuous native ungulates are white-tailed deer *(Odocoileus virginianus)* and, in some places, javelina *(Tayassu tajacu)*. Besides blackbuck, the most common exotics are axis deer *(Axis axis)*, fallow deer *(Dama dama)*, sika deer *(Cervus (Sika) nippon)*, nilgai antelope *(Boselaphus tragocamelus)*, aoudad *(Ammotragus lervia)*, and mouflon *(Ovis musimon)* or mouflon crosses. In spite of control measures, small predators such as coyote *(Canis latrans)*, gray fox *(Urocyon cinereoargenteus)*, and, occasionally, bobcat *(Lynx rufus)* remain. Nevertheless, no potential blackbuck predators were actually sighted during the blackbuck observations.

The study areas in Texas (Figures 6, 8a and b) cover gently rolling limestone country dissected in places into steep hills and valleys. All pastures had some form of brush clearance treatment over part of their area in the past. In most, grazing had probably been the only type of land use. Brush, including a high proportion of juniper bushes *(Juniperus ashei)* and occasional mesquite trees *(Prosopis glandulosa)* alternate with grassland scattered with "motts" of small oaks (mainly *Quercus virginiana virginiana*). The effect is that of a parkland. In all the small pastures, but also in some of the large pastures, supplemental feed in the form of pellets and hay was regularly offered to the wild and domestic animals.

Figure 6: Map of Texas with Edwards Plateau and locations of study sites (black dots).

There were no streams or rivers available to the animals in any of the areas where observations were concentrated. Ground water pumped by windmills ran into tanks, troughs, or ponds. Some of the pastures had man-made ponds which caught rain water. Annual precipitation varies greatly. In most years, potential evaporation exceeds precipitation, and droughts are characteristic. Severe windstorms and measurable snow are rare, but there are occasional hail or ice storms. Increased thunderstorm activity and cooler air moving in from the north contribute to the precipitation peaks in May and September. Flooding is common in fall.

During the summer the temperature difference between the days and nights is particularly noticeable, while mean annual temperatures remain fairly constant. Mean annual temperature is 19°C with a January mean of 9°C and a July mean of 28°C (Carr 1969). Nevertheless, temperatures can be high at any season. Temperatures often reach 32°C or above on summer days. During some winter nights, lows approach −18°C. Relative humidity is relatively uniform averaged over the whole year, but, like temperature, it fluctuates considerably during the day. Mean average humidity at 6:00 AM is 79%, but at 12:00 it is 52%, and by 18:00 it has dropped to 47% (Allison et al. 1975).

Wind speed averages 14 km/h. The prevailing direction is southerly except when disrupted by polar air masses which are common during the winter (Allison et al. 1975). These "northers" seldom last more than two to three days. Winter precipitation is mainly in the form of light rain or drizzle. Particularly during winter, mornings are frequently overcast, but skies usually clear by 10:00 hours.

Data on blackbuck behavior were gathered between 1971 and 1975 on a number of ranches on the Edwards Plateau, the principle ones being Greenwood Valley Ranch, Y. O. Ranch, South Fork Ranch, and Guajolote Ranch. The ranches were subdivided into fenced areas of native rangeland loosely referred to as "pastures." The pastures used for blackbuck observation fell into two groups: medium-size areas of 20 to 50 ha and large areas of 200 to 500 ha. Three pastures of each type were studied. Only one of them was used for hunting. Another large pasture had cattle on a rotation basis. All held a variety of native and exotic animals. There were generally 12 to 20 blackbuck in the medium-size pastures and 60 to 250 in the large pastures. However, one of the large pastures experienced a population crash during a particularly severe winter and subsequently approximated the medium-size pastures in blackbuck numbers.

The two major Indian study sites (Figures 7, 8c and d) contain the largest government-managed blackbuck populations in India. Only in areas of Rajasthan where the Bishnoi Community offers rigorous protection are there greater concentrations. One month each was spent observing at Velavadar Blackbuck National Park, Gujarat State, in the northwest (February 1980) and at Point Calimere Wild Life Sanctuary, Tamil Nadu, in the southeast (March 1980). Supplementary visits on one day each were made to a free-ranging population near Mudmal and a re-introduced population at the park Mahavir Harina Vanasthali both in central India's Andhra Pradesh and to the Guindy Park population around the governor's mansion in Madras, Tamil Nadu.

Velavadar Blackbuck National Park is in the Bhal grasslands about 70 km northwest of the port city of Bhavnagar. The Bhal forms a saline coastal strip along the Bay of Cambay in the Kathiawar Penninsula. Alternately baked in summer and drenched during the monsoon, the Bhal is characterized by extremes. Droughts may come as often as every second year, but cyclonic storms accompanied by incessant rains and flooding decimate blackbuck herds during some monsoons (Dharmakumarsinhji 1978). Since Velavadar is slightly lower than the surrounding Bhal, water rushes in and may stand knee-deep for 10 to 15 days. The waterlogged soil cracks as it dries. Then summer winds sweep the surface creating dust storms.

Figure 7: Map of India with major study sites (black dots): Velavadar Blackbuck National Park (north) and Point Calimere Wild Life Sanctuary (south).

The Bhal's three seasons are the monsoon (late June through October), winter (November through February) and summer (March through June). About 510 mm of rain may fall during the monsoon (Raychaudhuri et al. 1963). The grasses may grow to as much as a meter tall. Winter brings cold nights, cool mornings and pleasant days. Temperatures range from lows of −1° to 18°C to highs of 21° to 38°C. Summer temperatures climb to highs of 37° to 43°C (Raychaudhuri et al. 1963). Nights remain warm but calm.

About half of Velavadar Blackbuck National Park comprises areas of bare ground too salty for any vegetation except some patches of the low succulent *morad (Suaeda nudiflora)*. The rest of the park is covered in native grasses with an admixture of forbs. *Dharat (Cynodon dactylon)* is a common grass and the grass *jinjuva (Dichanthium annulatum)* has been shown in tests at nearby Dil Bahar to be highly palatable for blackbuck (Dharmakumarsinhji pers. comm.). A forb called *magamati (Phaseolus aconitifolius)* found at Velavadar is said locally to be the most preferred by both blackbuck and domestic grazers. The exotic tree *Prosopis juliflora,* promoted in the area about 12 years ago, is threatening to turn the Velavadar grasslands into thorn scrub. The Forest Department now hires workers to chop it out at the roots, but it still rings the park boundaries and roads.

Blackbuck are the only conspicuous native mammals in the park. Although officially banned, local domestic buffalo, cattle and goats wander in for grazing. Since there are no fences, blackbuck also wander out freely to glean what is left

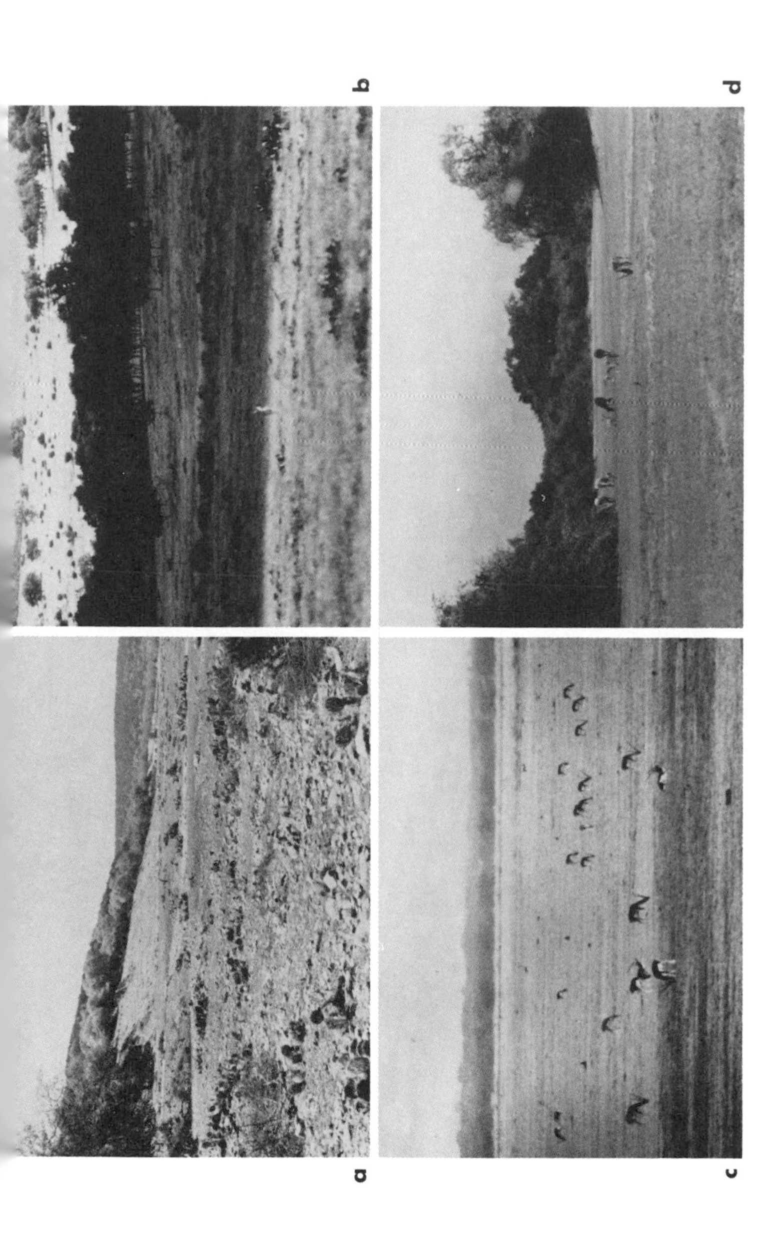

Figure 8: Blackbuck habitat in Texas and India. (a) Open valley among limestone hills at Greenwood Valley Ranch, Texas; a preferred area for territories. (b) Alternating grassland and trees in a pasture at South Fork Ranch, Texas, with a lone black-buck female going to fetch her fawn. (c) Flat, open grassland at Velavadar Blackbuck National Park, India, with a grazing bachelor herd. (Photos: E.C. Mungall.) (d) Opening in brush at Point Calimere Wild Life Sanctuary, India, with blackbuck grazing near domestic cattle. (Photo: S.E. Dougherty.)

in the surrounding wastelands and fields. Wolves *(Canis lupus)* hunt blackbuck as they occasionally pass through Velavadar (Shivbhadrasinhji pers. comm.), but jackals *(Canis aureus)* appear to be the main resident, native predator. The major pressure on the population comes from feral pariah dogs that make brazen hunting forays into the park every morning and evening.

By 1966, commercial meat hunting had reduced the blackbuck population using Velavadar from thousands to about 200 (Shivbhadrasinhji pers. comm.). In 1969 a sanctuary of 890 ha was established and blackbuck started to recover. M.K. Ranjitsinh counted approximately 2,000 blackbuck at Velavadar in 1974 (Shivbhadrasinhji pers. comm.). In 1976, the sanctuary was awarded national park status and enlarged to 1,783 ha. During study observations in February 1980, 1,624 ha of former buffer zone became part of the park. An unofficial staff census later that month tallied 2,006 blackbuck.

Point Calimere Wild Life Sanctuary on India's southeast coast with its gently rolling terrain and its patchwork of openings and brush clumps is much more difficult to census than are the flat, open plains at Velavadar. Recent counts have ranged from 506 to 1,108 within the 2,229 ha of forest and grassland habitat.

Point Calimere forms the seaward apex of the Cauvery River Delta and marks the juncture of the Bay of Bengal with the Palk Strait. Sri Lanka (formerly Ceylon) is only 45 km away. The coastal soils are deep with sand and are sandy loam to clay loam in texture (Karunakaran 1972). Dunes partially rooted by raja's demon (*Spinifex* sp.) and other plants line the sand beaches and help protect a belt of succulents, grasses and other low-growing plants. *Cyperus rotundus* is prominent and the grasses *Cynodon dactylon, Chloris barbata* and *Eriochloa procera* also occur. Several species of *Cassia, Zizyphus* and *Acacia* are conspicuous components of the brush clumps along with planted pandanus hedges and a thin zone of casuarina trees. The exotics *Opuntia* sp. (common in Texas) and *Prosopis juliflora* (the dominant tree at Velavadar) are invading but constitute only one component among many. The core of the reserved forest—nearly 1,500 ha of the sanctuary—is classified as tropical dry evergreen. Although the blackbuck avoid its thick scrub, they graze in its openings and make trips to its natural and man-made water holes. Sharing the sanctuary's grazing by a fee permit system are domestic buffalo, cattle and goats from local villages and feral ponies that run wild until wanted for sale. The so-called wild boar frequently sighted may also be feral and the shy complement of axis deer living in the forest were introduced (Krishnan 1971). Jackals were observed hunting blackbuck on several occasions, but no pariah dogs were noted inside the sanctuary. Indian wild dogs *(Cuon alpinus)* may occasionally hunt in the area.

Although Point Calimere has been noted as a blackbuck area since at least the 1800s (Jerdon 1874), its main notariety comes from the wealth of migratory birds, including thousands of flamingoes. The Forest Department has managed the area since 1892 with a major emphasis on providing grazing areas and producing forest products. More recently, a concern for limiting human disturbance to promote wildlife conservation culminated in the creation of "Point Calimere Wild Life Sanctuary" in 1967. Although the sanctuary has no fences, lack of suitable habitat deters much outside use by blackbuck.

In contrast to Velavadar, tropical Point Calimere has two rainy seasons each year. Rainfall varies both locally and yearly, and as much as 1,420 mm has been recorded in a single year. More than 60% comes during the northeast monsoon of October to December and about 30% during the southwest monsoon of May or June to September. The winds of June to August are strong enough to push sea water inland along the many broad, shallow swamp creeks which remain dry during the rest of the year. The shifting wind patterns in March to May and occasionally those in November create storms which can develop hurricane force (Karunakaran 1972).

Its cooling sea breezes earn Point Calimere the reputation as perhaps the coolest locality on the west coast of the Bay of Bengal. Maximum temperatures rarely exceed 37°C. The recorded minimum is 21°C. Mean daily temperatures rise from December and January lows of 22°C (minimum) until the peak in June. Humidity ranges from a December high of 82% to a July low of 70% for 8:30 and a December low of 68% to a June high of 77% for 17:30 (Karunakaran 1972).

In March, project members found mornings often foggy, with sunrise temperatures of 22° to 27°C. Humidity peaked at as much as 90% from 7:00 to 8:00, but stabilized at 60% by, at the latest, 10:30 after skies had cleared. It remained at this level until soon after sunset when it rose sharply. Midday highs ranged between 33° and 36°C.

MOUNTAIN GAZELLE

The mountain gazelle occurs from North Africa through the Near East into India. Throughout this large range, the numbers of these animals have been drastically reduced by hunting so that the mountain gazelle has been designated an endangered species. In Israel, however, the gazelles have been effectively protected by law since 1948. Due to this protection, their number increased, and in 1972, the population of mountain gazelle in Israel was estimated to total 3,500 to 4,000 animals (Mendelssohn 1974).

Obviously, the mountain gazelle is able to adapt to very diverse biotypes, including man-made ones. As its common name indicates, it differs from all other gazelle species in that its habitat includes very hilly and even rugged mountainous areas. On the other hand, mountain gazelle do not stand and walk on rocks as do the mountain ungulate specialists such as wild goats. Mountain gazelle are also found in plains areas, including croplands, and in the desert. They freely visit natural shrub forests as well as artificially planted pine forests and orchards where visibility under the trees is not severely restricted. They only are reluctant to enter areas with high ground vegetation (Mendelssohn 1974), as are all the species of the genus *Gazella*.

Because of this versatility, mountain gazelle can tolerate rather different climatic conditions. For example, in Upper Galilee, they live in areas with average temperatures of 6° to 8°C in winter (January) and with occasional subzero temperatures at night, and where the annual precipitation is 600 to 800 mm, occasionally reaching 1,000 mm (Mendelssohn 1972). Apparently, however, these climatic conditions are not optimal for mountain gazelle. The densest populations are found in areas which have an annual rainfall of only 200 to 400 mm and a yearly average temperature of about 21°C, whereas in Upper Galilee, the yearly average is 15°C (Mendelssohn 1974).

The mountain gazelle in Israel are resident and do not migrate with changing seasons in recent times. The southern part of their habitat overlaps with that of dorcas gazelle. In certain areas, such as En Gedi, they may occasionally meet Nubian ibex *(Capra ibex nubiana)* and in others, such as the valleys in the Golan Heights, wild boar *(Sus scrofa)*. Natural predation on mountain gazelle in Israel is rare although populations in the Negev are exposed to low populations of potential predators such as wolf, caracal *(Felis caracal)*, and leopard *(Panthera pardus)*. However, mortality due to human activity has greatly influenced the gazelle populations. In Israel, they have been reduced by legal hunting in the past and by poaching in recent years. They also suffer mortality from rodenticides (thallium sulphate, fluoracetamide) and organochlorine insecticides (Mendelssohn 1974).

Field observations on behavior of mountain gazelle were conducted on two study areas in Israel from August 1970 to June 1972 (Figures 9, 10). In the Northern Negev, a study area of about 40 km² was located in the gently rolling loess soil of the Philistian plains along the Gaza Strip about 40 km west of Beersheba. This area, with an annual rainfall of 150 to 200 mm (Zohary 1962), included the kibbutzim (collective farms) Nir Oz, Nirim, and En Hashalosha.

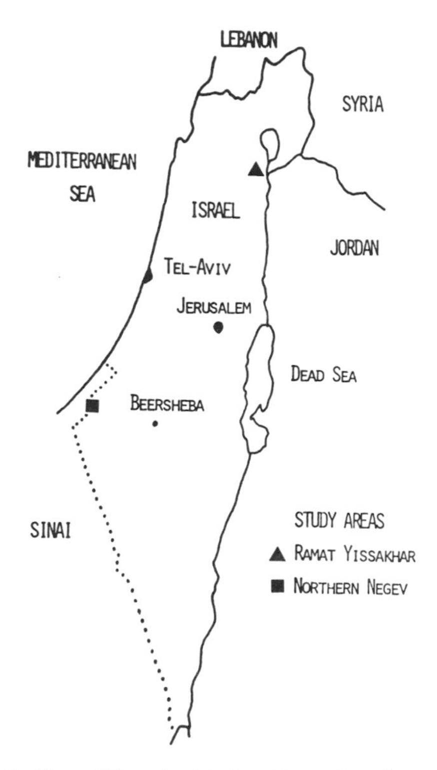

Figure 9: Map of Israel with locations of study areas.

Figure 10: Mountain gazelle habitat in Israel. (a) General view of the agricultural habitat in the northern Negev with two kibbutzim in the background (aerial view). (b) Wheat stubble in the northern Negev with a kibbutz in the background and a gazelle herd crossing the wheat stubble (aerial view). (c) General relief of the Ramat Yissakhar terrain with the Jordan River Valley in the background (aerial view). (d) Rocks and shrubs at Ramat Yissakhar with two gazelles in the foreground. (Photos: G.A. Grau.)

The relatively flat terrain, about 100 m above sea level, was cultivated, mostly in small grains during the wet season (December-March), and fallow during the dry months. Average monthly temperatures were moderated by the Mediterranean Sea and ranged from 12°C in February to 25°C in August (Zohary 1962). Several citrus orchards were situated on the study area. Gazelle density in this area was relatively low (2.5 per km^2), but observation was quite easy due to the flat terrain.

The other study area of about 10 km^2 was located in Nahal Yissakhar, a deep, wide *wadi* (dry creek) crossing the southern part of the Ramat Yissakhar region in the Lower Galilee. The vegetation consisted of dwarf-shrub communities within the *Zizyphus lotus - Zizyphus spina-christi* association characterizing the steppe of the Irano-Turanian vegetation type. This area received an annual rainfall (December-March) of 350 mm and ranged in elevation from 200 m below sea level in the valley to 300 m above sea level on the northern slopes. The average monthly temperature ranged from 12°C in January to 30°C in late August (Zohary 1962). Cattle grazed in the valley all year long, and the surrounding area was cultivated and planted in small grain crops. Gazelle density reached 37 per km^2 (Baharav 1974), but observation was difficult due to the rough terrain.

THOMSON'S AND GRANT'S GAZELLES

Thomson's gazelle, or "tommy" as this species is usually called in East Africa, inhabit a relatively small range in northeastern Tanzania and southeastern Kenya. A subspecies, *Gazella thomsoni albonotata,* occurs in eastern Ethiopia to southern Sudan (Haltenorth 1963). Thomson's gazelle show a strong preference for open short-grass plains. In the Serengeti area, they migrate to and through the *Acacia-Commiphora* woodlands in the northern half of the national park during the dry season. During these migrations, they must cross high-grass plains, but they do not stay there. Within its limited habitat, the tommy is very abundant. In 1965, the population in the Serengeti area was estimated at 600,000 (game warden report for the annual meeting of the Serengeti Research Council). No accurate census has been conducted, but later estimations by several authors vary considerably. For example, Hendrichs (1970) spoke of more than 900,000 Thomson's gazelle in Serengeti, whereas Schaller (1972) estimated them to be only 200,000. In any case, these differences reflect the different opinions of the authors, but they do not reflect observed increases or decreases in population size.

The considerably larger Grant's gazelle ranges from Ethiopia to the middle of Tanzania, and from Lake Victoria to the coast of the Indian Ocean. Thus, its habitat is far more extensive than that of the smaller Thomson's gazelle which lives together with Grant's gazelle in Serengeti. Here, the game wardens estimated the total number of Grant's gazelle to be about 40,000 in 1965. As with Thomson's gazelle, other authors gave rather different figures. For example, Schaller (1972) estimated the number of Grant's gazelle in the Serengeti National Park to be about 10,000, and Hendrichs (1970) spoke of only somewhat more than 3,000. This last figure appears to be a severe underestimation. However, it is certain that Grant's gazelle are less numerous than Thomson's gazelle in the Serengeti area. The Grant's gazelle inhabit relatively open woodland (particularly areas with scattered whistling acacias) as well as open plains and may even be found in semi-arid areas. Although they also prefer short-grass plains, they do not avoid high-grass areas as strictly as do Thomson's gazelle.

Both Grant's and Thomson's gazelles migrate between the open plains and the woodlands in Serengeti. Roughly speaking, the gazelle leave the woodland with the beginning of the small rains in November and migrate toward open plains in the southeast. They stay there, moving between Lake Ndutu and Lemuta Hill, until the end of the long rains in May. Then, they cross the Central Serengeti Plains in a northwestern direction and reach the junction zone (e.g., at Seronera or at the Mbalangeti Gap) at the end of May or the beginning of June. Later, they stay in the so-called "corridor" of the Serengeti National Park (the portion nearing Lake Victoria), or in the Grumeti, Ikoma, and Mara areas, until the onset of the small rains (Grzimek and Grzimek 1960).

This rather general picture needs additional details. First, not all the Thomson's and Grant's gazelles participate in the emigration to the open plains during the rainy season. A certain percentage remains in the woodland distributed on several *mbuga* (natural clearings in the woodland). Movements on the plains depend very much on the extent of the dry season in January and February between the small and the long rains. If the country becomes very dry during this time and/or the long rains start late, the herds may come back to the junction zone as early as February or March and may emigrate once more when the long rains start (in March).

Furthermore, at the beginning of the dry season after the long rains, the regular and predictable migration of the gazelles is over after the herds arrive at the junction zone. What happens later depends on the extent of the drought, the distribution of high and short grass on the *mbuga* in the woodland, the extent of local burning, and rare local rain showers. In general, the Thomson's gazelle go deeper into the woodland during the period from July to September, but the distances and the directions of these movements depend on the previously mentioned conditions which vary from year to year.

Some of the Grant's gazelle migrate more or less together with the Thomson's gazelle. However, there are Grant's gazelle which do not participate in this migration but remain in the plains during the dry season. Obviously, Grant's gazelle are less water-dependent than Thomson's gazelle and all the other East African plains ungulates, except the oryx antelope *(Oryx beisa callotis)*. The Grant's gazelle which remain in the open plains, gather around the Gol, Barafu, and Simba Kopjes or at Naabi Hill in May or June. They stay there in large numbers until August/September. In late September, they start moving toward the woodland to the northwest, following the *korongo* (creeks which have water only during the rainy season). In October, these herds arrive in the areas around Lake Magadi and at Seronera. By this time, however, the Thomson's gazelle and those Grant's gazelle which had migrated with them, are usually already on their way back from the woodland to the junction zone where they "wait" for the rains to fall in the open plains. As soon as the small rains start in November, the gazelle emigrate again to the plains to the southeast.

In the plains during the wet season, and in the woodland during the dry season, the gazelle are together with a great many other ungulate species such as wildebeest *(Connochaetes taurinus)*, plains zebra *(Equus quagga)*, eland antelope *(Taurotragus oryx)*, topi *(Damaliscus lunatus topi)*, kongoni *(Alcelaphus buselaphus cokei)*, impala, waterbuck *(Kobus defassa)*, reedbuck *(Redunca redunca)*, and giraffe *(Giraffa camelopardalis)*. Grant's and Thomson's gazelles show a certain social attraction to each other, but not to the other game animals (occasional exceptions are the territorial individuals, see p. 190) Of course, the gazelle do not avoid the other ungulate species. For example, they may locally and temporarily graze together with other ungulates in close proximity. However, such congregations are only brought about by environmental factors; a locality may have favorable feeding conditions or trees which give shade on hot days, etc. Also, there usually are neither interactions with the other species, nor any reactions to or synchronizations with the other species' activities. Thus, animals of other species are not much more than inanimate objects to the gazelles. Only when large zebra herds and, above all, big wildebeest concentrations, move in, do the gazelle react to them by leaving the area.

Many predators prey upon Thomson's and Grant's gazelles in the Serengeti area. Lion *(Panthera leo)*, leopard, and spotted hyena *(Crocuta crocuta)* take their toll. Cheetah *(Acionyx jubatus)* and wild dog *(Lycaon pictus)* are great gazelle hunters. Golden jackal, blackbacked jackal *(Canis mesomelas)*, baboon *(Papio anubis)*, and tawny eagle *(Aquila rapax)* prey upon fawns. These are only the major predators. The list of those which may occasionally take a gazelle and particularly a fawn, is considerably longer, and it comprises any mammalian predator up from the size of a cat and any larger bird of prey as well as several big reptiles such as crocodile *(Crocodylus niloticus)* or python *(Python sebae)*. Of course, some of these predators (e.g., leopard) are not particularly abundant and are more or less restricted to a few circumscribed localities; however, others (e.g., hyena and lion) are quite numerous and may occur almost anywhere in Serengeti. Thus, one could expect the gazelles to spend their life in permanent alertness and fear. However, this is not the case. On the contrary, it is surprising how little they care about the numerous predators and how peacefully they live and behave in spite of their presence. Even direct and successful hunting actions of the predators usually cause no long lasting disturbances in the herds. Peace returns and life goes on soon after the predator has made its kill.

Almost the whole Serengeti ecosystem is situated between 1,200 m and 1,800 m above sea level, and monthly averages of maximum daily temperatures fluctuate between 25° and 32°C (in shade), minimum daily temperatures vary between 11° and 18°C (Kruuk 1972) with June and July being the coolest months. Daily (shade) temperature differences of 15° to 20°C between the low before daybreak (about 6:00) and the hottest time in the early afternoon (14:00-15:00) are quite common.

In a "normal" year, the "small rains" fall in November and December. January and February are dry with only occasional showers. The "long rains" last from March through May, followed by a long dry season from June to October. However, there are great yearly variations in the onset, the end, and the intensity of the rains. The average annual rainfall measured at Banagi between 1937 and 1959, was 772 mm with variations from 466 mm to 1,074 mm (Grzimek and Grzimek 1960). These figures are fairly typical of the woodland areas in the Serengeti. The average rainfall in the southeastern plains is below 500 mm per annum (Kruuk 1972). Relative humidity ranges from a mean monthly minimum of about 15% during the dry season to about 40% during the rains; the mean monthly maximum is about 85% (Schaller 1972). As with temperature, humidity often varies considerably in the course of a day. For example, in the Togoro plains on 10 days in May 1965, the average varied by 45% with its maximum in the hours before daybreak and its minimum at mid-afternoon (Walther 1973a). The prevailing winds come from east to southeast, and they can reach high velocity particularly in the plains. For instance, in the Togoro plains in May 1965, the maximum speed measured was 10 m/sec (Walther 1973a). Usually, it blows least between 21:00 and 6:00. The wind freshens up in the morning hours and remains more or less constant until about 12:00, then it declines.

The southeastern part of the Serengeti ecosystem, i.e., roughly the area between Ngare Nanyuki in the north, the Ngorongoro Crater in the south, the Simba Kopjes in the west, and the Salai Plains in the east—consists of gently rolling short-grass plains. Grass species of the genera *Sporobolus, Eragrostis, Digitaria,* and others are common, and forbs such as *Solanum incanum* and *Indigofera basiflora* are locally abundant. (The interested reader may find more detailed information on the vegetation of the Serengeti National Park, including that of high-grass areas and of woodlands, in a number of recent books and publications, e.g., Grzimek and Grzimek 1960, Schaller 1972, Kruuk 1972, Herlocker 1975, Sinclair and Norton-Griffiths 1979.) Locally there are stair-like erosions, and several groups of large rock outcrops *(kopjes)* are interspersed in the open country. Several long creeks *(korongo)* intersect the area. During the dry season, these are either completely dry, or else some water remains at their deepest points, depending on the extent of the drought. At several places, bushes and acacia trees border the *korongo.* Forest begins northeast of the Barafu Kopjes, i.e., in the ecotone of the Ngare Nanyuki area.

During the wet season, almost all the big game species, typical of the East African plains, are found in the area under discussion. In addition to Thomson's and Grant's gazelles, zebra, wildebeest, and eland antelope often are seen in great numbers. In the dry season, all the wildebeest and zebra, and most of the eland and Thomson's gazelle leave the area. Grant's gazelle remain in large numbers, and a few tommy stay with them.

South and west of the described short-grass areas, there is a broad transitional zone with grasses of 1 m and more tall. This transitional area is intensively grazed mainly by wildebeest and zebra during their movements in the wet season. During the dry season, burning is common, and most animals leave this area. Thomson's gazelle only cross it during migration. Grant's gazelle may remain there somewhat longer.

An open woodland with chains of quite remarkable hills covers much of the northern and western sectors of the Serengeti National Park. The tree communities vary from the tall gallery forests, locally characterized by the striking fever tree, along the rivers to the typical Serengeti woodlands which are dominated by *Commiphora* and several *Acacia* species. "*Mbuga*" (natural clearings) of varying size, from less than 1 km^2 to about 20 km^2, are interspersed in the woodland. These *mbuga* are high-grass or short-grass areas with basically the

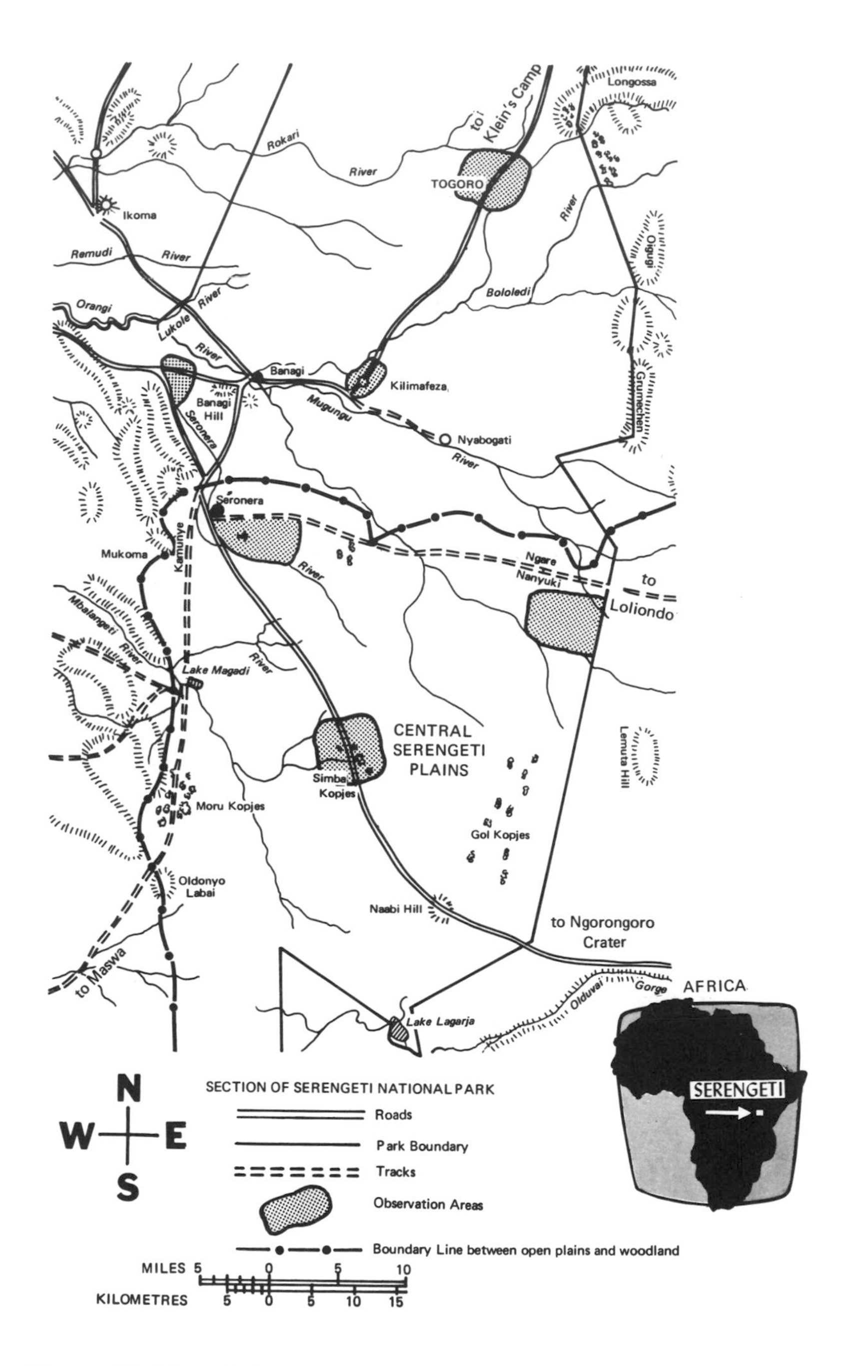

Figure 11: Map of the south-eastern section of Serengeti National Park, Tanzania, East Africa, with locations of study sites. (The woodland begins north and west of the demarcation line.)

same grasses and forbs as are found on the open plains. They are frequently framed by broad belts of scattered whistling acacias.

Some ungulate species, such as impala, waterbuck, and topi are found in the woodlands all the year round. For most of the plains animals, the woodlands form the dry season refuge. Gazelle only cross the forests during migration, but they stay for considerable periods on the *mbuga* and in the areas with scattered whistling acacias. Particularly during the second half of the dry season, burning is frequent in the woodlands, and then the feeding conditions become poor there.

In addition to a short-term study in the Ngorongoro Crater in January 1964 (Walther 1964a, 1965), data on behavior of Grant's and Thomson's gazelles were gathered in the Serengeti National Park from January 1965 through December 1966, in July and August 1972, and from May 1974 through April 1975. Several study areas were used, widely distributed in the park (Figures 11, 12).

Figure 12: Thomson's and Grant's gazelle habitat in Serengeti. (a) Open plains at Ngare Nanyuki with a herd of Grant's gazelle in file in front of the single trees. (b) Open plains with rock outcrops (Simba Kopjes) in the background and Grant's gazelle in the foreground. (c) *Mbuga* (Togoro) in woodland with Thomson's and Grant's gazelles. (d) Junction zone (Seronera) with migratory herds of Thomson's gazelle. (Photos: F.R. Walther.)

These were predominantly the Togoro Plains (mainly in 1965/66 when this area was almost untouched, whereas it was largely destroyed by road building, gravel pits, and fencing for ecological research in later years), the Musabi Plains, the area between Ngare Nanyuki and the Barafu Kopjes, the area of the Gol Kopjes, the area around the Simba Kopjes, the areas near Seronera, (west of) Banagi (where the "old corridor road" hits the "new corridor road"), and at Kilimafeza. Musabi, Togoro, Kilimafeza, and the area west of Banagi were *mbuga* in the woodlands. The observation area in the Togoro Plains covered about 5 km^2. The entire Togoro Plains approximated 10 to 15 km^2. (Names of such locations are often treated "generously" in Africa, and, thus, it is often not precisely determined where the named area begins and where it ends.) Kilimafeza and Banagi were smaller *mbuga* of about 1 to 3 km^2. None of these places were independent ecological units. Togoro came closest because it was surrounded by relatively dense woodlands which extended for miles on all sides and, with the exception of the migratory periods, gazelles were not found in these woodlands. Simba Kopjes, Gol Kopjes, and the area between Ngare Nanyuki and Barafu Kopjes were open plains areas interspersed with the described rock outcrops and intersected by *korongo*. To the north, the Ngare Nanyuki area was bordered by a gallery forest along such a *korongo*. Seronera was a junction zone where open plains and woodland met.

3

The Place of Territoriality Within the Social System

GREGARIOUSNESS AND ISOLATION TENDENCY

Basically, the gazelle and their relatives are gregarious animals; however, isolation tendencies temporarily occur in individuals under certain circumstances. When isolation tendencies are not in effect, these animals form groups ranging from two to hundreds and even thousands of individuals. The differences in herd size depend on species (e.g., gerenuk and dibatag usually form only small groups of less than 10 animals), on season (the largest herds occur during migration periods), on environment (generally, the herds are larger in open plains than in forest areas), and on population density. Population density may even have effects upon the group composition. For example, when the mountain gazelle population in Israel was down to its lowest level in the years before they were protected by law, these animals were usually seen singly or in pairs (Mendelssohn pers. comm.). After the population had somewhat increased due to protection, all-male groups as well as female groups of several members were observed. However, even nowadays, mixed herds (i.e., herds composed of several adult males and several adult females) are not seen in Israel (Grau 1974), but they probably did occur there in former times when the population was large. Reports from the beginning of this century speak of herds of 500 to 600 gazelles in Israel (Mendelssohn 1974), and, after all we know from other gazelle species, such big herds are always mixed.

In principle, the Antilopinae herds are open societies, i.e., their members can come and go as they like. Frequently this openness goes so far that one can only say these animals live in herds, but that it is hard to speak of a definite herd because there is continual splitting and amalgamating. However, it is not always so, and sometimes herds may remain constant in size and composition over weeks and months. In spite of their gregariousness, gazelles and their relatives are not "contact animals" but are rather "distance animals" which keep a certain "individual distance" (Hediger 1941) between each other. This individidual distance may vary with sex, age, and activity (Walther 1977a) but usually does not come down to zero, i.e., to bodily touch.

The solitary status of an individual is always temporary in gazelles and their relatives. Sometimes an animal may be merely accidentally separated from its group. For example, hunting actions of predators or aggression by a superior and very intolerant conspecific may cause an individual to split from the group. Probably, accidental separation is also true for quite a number of cases in which non-territorial adult males are wandering around on their own. In other cases, such solitary wandering adult males may be owners of territories who temporarily have left their territories. Or, they may be individuals which have left their groups and are now searching for a location suitable to establish a territory. In any case, the solitary status of non-territorial individuals is a rather temporary matter, and they cannot be considered to be a special social class in these species.

Genuine isolation tendencies occur only in three instances in the Antilopinae. During the first weeks of its life, a fawn moves away from its mother after it has been nursed and cleaned, in order to "lie out," i.e., it walks away, beds down and remains separated from its mother for hours until she calls it for the next nursing. Secondly, a female may separate from the herd when giving birth. This period may also include some time before parturition and a few days after it. The third case of isolation tendency involves territorial bucks. They leave the herds in order to establish their territories and to stay there.

THE SOCIAL GROUPS

As in many wild ungulates, at least three basic social units, besides territorial males, can be distinguished in Antilopinae: female groups, all-male (or bachelor) groups, and mixed herds. In addition, one may speak of harem groups or, at least, pseudo-harems in certain cases. Figure 13 summarizes the most important aspects of social organization in Grant's gazelle that can be applied with minor variations to most of the other Antilopinae species.

Female Groups

As mentioned above, a female may isolate herself before giving birth and remain solitary in the vicinity of her out-lying fawn. Thus, in the extreme case, a female group may consist of only one female or a mother with her fawn. Most of the female groups, however, comprise more members. They consist of females of all age classes with or without offspring. Adolescent males may still be found in female groups, but no males beyond this stage. A special case of a female herd may arise when mothers with very young fawns join exclusively with each other after each of them has gone through an isolation period after giving birth. One may speak of a mothers group in such a case.

The size of female groups (Figure 14) partly depends on population size and density. Female groups of 2 to 20 individuals are frequent, and larger groups occasionally occur in all the more gregarious species under discussion. In Grant's gazelle (Walther 1972a) as well as in mountain gazelle (Grau 1974), the

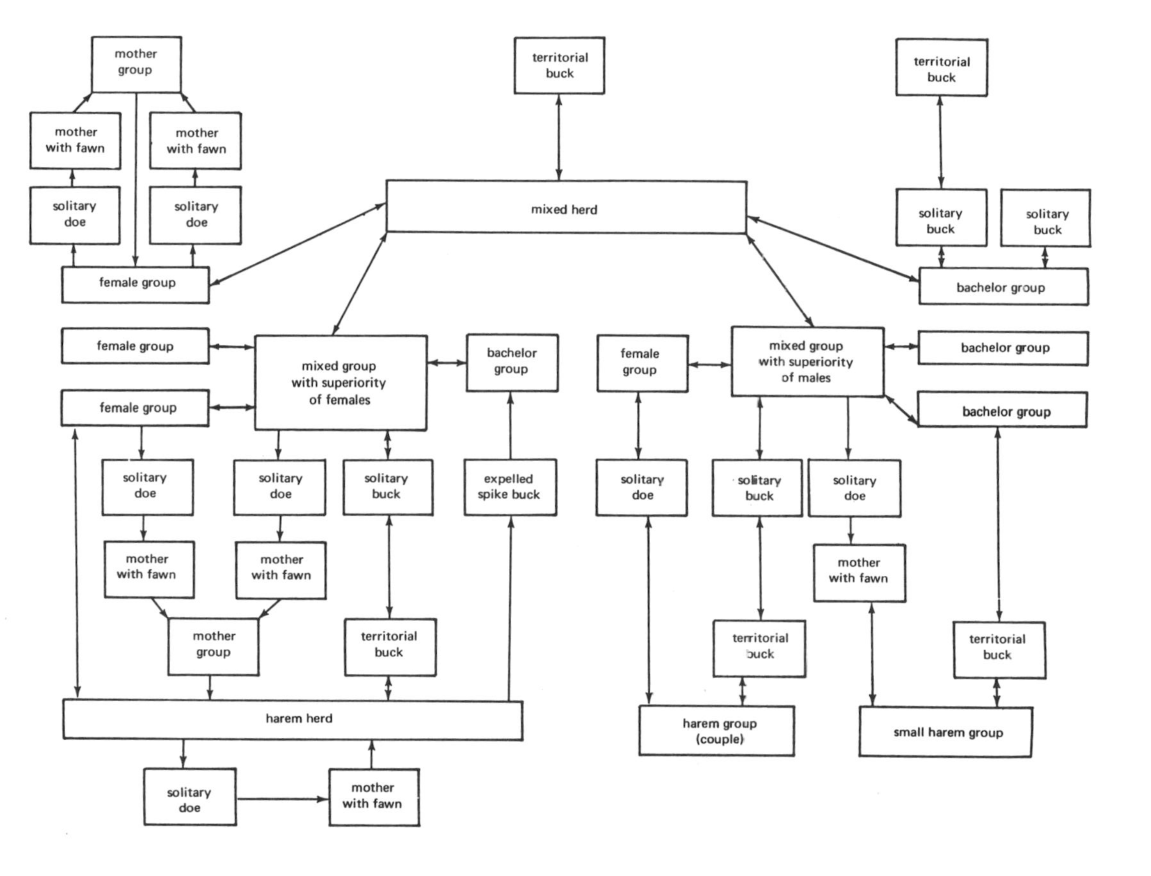

Figure 13: Scheme on social organization in Grant's gazelle.

maximum size of a female group is about 40 members. In blackbuck, it may range up to 80 (Cary 1976b). In Thomson's gazelle within its abundant Serengeti population, female herds can even reach the size of about 400 animals. Two factors make the occurrence of large female herds rare even in an abundant population. The larger a herd is, the easier it splits into smaller groups, and the larger a female herd is, the sooner it is joined by males. In the latter case, of course, it is no longer a female herd but becomes a mixed herd.

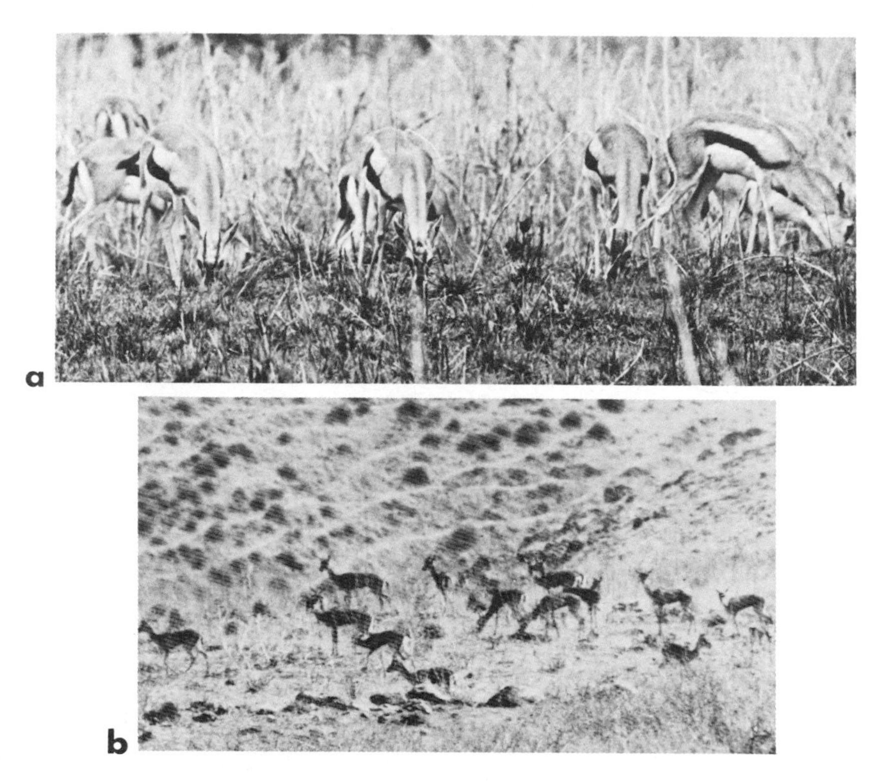

Figure 14: Female groups. (a) Thomson's gazelle. (Photo: F.R. Walther—Serengeti National Park, Tanzania.) (b) Mountain gazelle. (Photo: G.A. Grau—Ramat Yissakhar, Israel.)

All-Male Groups

All-male groups (Figures 4c, 15) or bachelor groups, comprise males from adolescence to very old age. Adolescent males sooner or later leave the female herds and join the all-male groups. Thus, they, as well as subadult males, are frequently found in bachelor herds. However, at least in Antilopinae, it is not true that all-male groups are generally composed of such immature males as has been presumed in literature (e.g., Ewer 1968). On the contrary, adult males are a constitutive element of all-male groups and make up a considerable proportion (Walther 1972a), perhaps with the exception of gerenuk (Leuthold 1978a). Brooks (1961) mentioned subgroups of immature males within large bachelor herds of Thomson's gazelle. However, such subgroups of young males do not frequently occur in this and other species. In Indian blackbuck, individual bonds between two immature males (Mungall 1978b) appear to be more common and important than the formation of large age class groups within bachelor herds.

Figure 15: All-male groups. (a) Grant's gazelle. (b) Thomson's gazelle. (Photos: F.R. Walther—Serengeti National Park, Tanzania.)

The range of group size is about the same as in females; however, the majority of the male groups tend to be toward the lower end of the scale. Occasionally, all-male groups of 80 and more bucks can be seen in Grant's and Thomson's gazelles and in blackbuck. However, the average all-male groups are usually somewhat smaller than female herds in gazelles (Estes 1967, Walther 1972a,

Mungall 1978a). On the other hand, in a study of blackbuck at a plains site in India, female groups never reached the average size of 10 members found in other areas whereas all-male groups of more than 20 were common and all-male groups of more than 80 were by no means exceptional. As with female herds, large bachelor groups are easily joined by members of the opposite sex, and thus, they are converted into mixed herds.

Mixed Herds

Mixed herds (Figure 16) consist of several males and females. From the minimum of three animals (two males and one female), they range up to several hundreds, e.g., to more than 400 in Grant's gazelle (Walther 1972a), and in very sizable populations, such as tommy in Serengeti and blackbuck in India in the past (Jerdon 1874), to thousands of animals. Thus, their average size is considerably larger than that of bachelor or female groups, particularly in open plains areas (Walther 1972a). Mixed herds frequently occur during migration periods but are not restricted to these times. Immatures and juveniles, but not fawns younger than half-grown, are found in mixed herds. Adult males and females make up the greatest portion of the animals within the mixed herds and are their constitutive elements (Walther 1972a). Relatively frequently, the sexes form subgroups within the mixed herds.

Apparently, mixed herds only occur in Antilopinae when the population has reached a certain density. In our investigations, they were frequent in Thomson's and Grant's gazelle. In blackbuck, the absence of mixed herds on some Texas ranches seems to be due to small population size. A pasture with about 125 blackbuck had mixed herds of as many as 60 animals. Large Indian populations always included mixed groups. No mixed groups were observed in the recent studies on mountain gazelle—probably because the population in Israel has not yet reached an optimal size after its drastic decline in the past. Thus, it is unlikely that the observed absence of mixed herds in mountain gazelle indicates a principle difference from the social organization of the other species under discussion. In springbok, mixed herds are quite common (Bigalke 1972). In gerenuk, a species whose group size generally is considerably smaller than that of most of the other Antilopinae species, the situation appears to be somewhat different. All adult males seemed to be territorial in Leuthold's (1978a) observation area. Thus, mixed groups with two (not more) adult males were only exceptionally observed.

Harems and Pseudo-Harems

The term "harem group" has special problems when applied to gazelles and their relatives. It implies that a stable group of females is together with one male for a long enough time that, at least theoretically, this male can service most, if not all, of the females. Among the four species under discussion, such harem groups exist for certain only in Grant's gazelle, but even here only under certain conditions (Walther 1972a). Brooks (1961) spoke of harem groups in Thomson's gazelle, and he even considered them to be particularly important in this species. However, in tommy as well as in other Antilopinae species, the female herds usually only visit a territorial male in his territory, stay with him, at most, for a few hours, and then leave again. Thus, the territorial male and the female group remain two separate and independent social units. When there are many neighboring territories, a male may be seen together with a given female herd throughout its daily circuit; however, this is not the same male but always another one as the females pass from one territory to the next.

Figure 16: Mixed herds. (a) Thomson's gazelle (alarmed). (b) Grant's gazelle. (Photos: F.R. Walther—Serengeti National Park, Tanzania.)

Obviously, Brooks (1961) has not been aware of these changes in the "harem" males. In short, these female groups are not necessarily stable units themselves; they certainly do not stay exclusively with one male, and they do not stay long enough with any male for him to service more than one or two of the females. Thus, the term "harem" in its strict sense cannot be applied. One could possibly speak of a "pseudo-harem" (Figure 17a) with respect to such a female herd staying together with a territorial male for a short time.

In Grant's gazelle, particularly on the *mbuga* in the woodland, a group of females may sometimes remain in the territory of a buck for several weeks and months—possibly because the home range of this female group more or less coincides with the territory of this male. This can be said to be a harem (Figure 17b) and territorial behavior is combined with harem behavior in such a case (Table 1). Harem formation without a combination with territorial behavior of the male, as for example in plains zebra (Klingel 1967), does not exist in Antilopinae. The home ranges of mountain gazelle female groups more or less coincide with the male's territory. But these females occasionally leave the territories and the composition of the group may change from time to time. The situation seems to be similar in gerenuk (Leuthold 1978a).

Occasionally, a pair (Figure 17c), i.e., one male with one female, can be seen in Antilopinae. Sometimes an isolated mother with a fawn lying out in the vicinity may join or be joined by a male in his territory. A territorial male may find an estrous female in his harem or pseudo-harem. Then he and his favorite may temporarily separate from the others during the course of the courtship ritual. A female in estrus and intensively courted by a territorial male may linger a while with him after the other females leave his territory. In short, pairs are not social units of their own in the Antilopinae species under discussion. Instead, they usually are special cases of harems or pseudo-harems.

Since the harems or pseudo-harems are female groups temporarily staying with a territorial male, their size frequently is the same as that of pure female groups. About as frequently, however, their size somewhat differs from that of pure female groups. For example, in Thomson's gazelle, the female herds frequently are relatively large in Serengeti, but the territories of the males are comparatively small and often close together. When a large female herd enters such a territorial mosaic, each buck tries to herd a group of females in his territory, and he cuts out a section of the big herd. Thus, these pseudo-harems are often smaller than many of the "untouched" female herds. On the other hand, when the territories of the males are large and not as close together, and the female groups are relatively small and stay with a territorial male for a longer period—as it is with Grant's gazelle in the *mbuga* situation—a territorial buck may gather and keep several female groups in his territory as they arrive one after the other in the course of time. Then, the average size of such a harem can be greater than that of the (pure) female groups in the area.

THE TERRITORIAL MALES

Territoriality and Social Grouping

As already mentioned, only adult males become territorial in Antilopinae

Figure 17: Harems and pseudo-harems. (a) Territorial tommy buck with female group entering his territory (pseudo-harem.) (b) Harem group of Grant's gazelle on a *mbuga* (observed for a period of about five months). (c) Pair of Grant's gazelle. This buck had been with a bachelor group, but he became territorial when a single doe gave birth in the vicinity and remained there with her neonate for several days. (Photos: F.R. Walther—Serengeti National Park, Tanzania.)

Table 1: Records on a Territorial Buck of Grant's Gazelle and His Harem in Togoro in 1965

		ad♂	ao♂	ad♀	sa♀	ao♀	hgf	faw	nfa	Total	Comments
Jan.	8	1	0	4	0	0	0	0	0	5	
	13	1	0	3	0	0	0	0	0	4	
	18	1	0	7	0	2	0	0	3	13	
	28	1	0	6	0	2	0	3	0	12	3 faw = 3 nfa on Jan. 18
Feb.	7	1	0	7	0	2	0	3	1	14	
	14	1	0	7	0	2	0	4	0	14	1 faw = nfa on Feb. 7
	23	1	0	7	0	2	0	4	0	14	
	27	1	0	7	0	2	0	3	0	13	1 fawn killed by lions
Mar.	3	1	1	13	0	2	0	3	0	20	
	10	1	1	10	0	2	0	3	0	17	
	22	1	1	14	0	2	0	3	2	23	
	30	1	1	14	0	2	0	3	1	22	1 nfa killed by jackals
Apr.	9	1	1	14	0	2	0	4	0	22	1 faw = nfa on Mar. 30
	13	1	0	4	0	0	0	1	0	6	After hunting by wild dogs
	19	1	1	13	0	2	0	4	0	21	1 ad♀ killed (by wild dogs?)
	24	1	1	13	0	2	0	4	0	21	
May	3	1	1	10	2	0	0	4	0	18	2 sa♀♀ = 2 ao♀♀ until Apr. 24
	10	1	0	10	2	0	0	4	0	17	ao♂ chased away by Pasha
	17	1	0	10	1	0	0	4	0	16	1 sa♀ killed by cheetah
	27	1	1	21	3	3	0	6	0	32	
June	1	1	0	17	1	0	0	4	0	23	
	7	1	0	14	1	0	0	4	0	20	
	16	1	0	17	1	0	4	0	0	23	4 hgf = 4 faw until June 7
	21	1	0	13	1	0	3	0	0	18	
July	1	1	0	12	1	0	3	0	0	17	
	5	1	0	13	1	0	3	0	0	18	
	16	1	0	13	1	0	3	0	0	18	
	24	1	0	5	0	0	0	0	0	6	
Aug.	4	1	0	4	0	0	0	0	0	5	
	8	1	0	6	0	1	0	0	0	8	
	18	1	0	1	0	0	0	0	1	3	
	30	1	0	0	0	0	0	0	0	1	
Sep.	7	1	0	1	0	0	0	1	0	3	
	16	1	0	2	0	0	2	0	0	5	
	21	0	0	0	0	0	0	0	0	0	
	30	0	0	0	0	0	0	0	0	0	
Oct.	9	0	0	0	0	0	0	0	0	0	
	14	1	0	4	0	0	0	0	0	5	
	19	0	0	0	0	0	0	0	0	0	
	22	0	0	0	0	0	0	0	0	0	
Nov.	2	1	0	0	0	0	0	0	0	1	
	8	1	0	2	0	0	0	0	0	3	
	11	1	0	5	0	0	1	0	0	7	
	23	0	0	0	0	0	0	0	0	0	
Dec.	3	0	0	0	0	0	0	0	0	0	
	19	1	0	3	0	0	0	0	0	4	
	23	1	0	5	1	0	0	0	0	7	
	29	1	1	6	0	0	0	0	0	9	

Signs and abbreviations: ad♂ = adult male (= the territorial buck named "Pasha"); ao♂ = adolescent male; ad♀ = adult female(s); sa♀ = subadult female(s); ao♀ = adolescent female(s); hfg = half grown fawn(s); faw - fawn(s); nfa = neonate fawn(s)

(Figure 18). Up to the adolescent age, a male is a member of a female group and later a member of an all-male group or mixed herd. In mountain gazelle, territoriality is continuous, i.e., a male establishes his territory and stays in it, at least under the present population density conditions in Israel. Likewise, territoriality seems to be a rather continuous condition in gerenuk (Leuthold 1978a). In the other Antilopinae species, the adult males do not usually keep their territories for years without interruption. Instead, territorial periods and non-territorial periods alternate with each other during the life of the individual and frequently within the course of the same year. During non-territorial periods, the adult males form or join all-male groups and mixed herds. Thus, the mixed herds and the all-male groups are the social background from which the males become territorial and to which they return again and again. Furthermore, not all adult males become territorial at the same time. Consequently, males in their territories and adult males in bachelor groups or in mixed herds are found beside each other throughout the year (again apparently with the exception of the gerenuk, at least in certain areas—p. 193).

Seasonal Changes in Territoriality

In principle, territoriality is not bound to a definite season in these animals. An adult male may establish a territory, keep it, and abandon it at any time of the year. However, there are seasonal peaks in territoriality, i.e., seasons in which relatively many males become territorial and/or hold territories, alternate with seasons in which the numbers of territorial individuals are comparatively low. These highs and lows in territoriality are partly linked to reproductive behavior. However, reproduction is not limited to a definite season either, but also shows seasonal increases and declines in most of the Antilopinae species. For example, the peak of the mating goes from September through November in mountain gazelle in Israel (Grau 1974); however, this does not have much effect on territoriality which is a continuous phenomenon in this species (see above). In blackbuck, increased reproductive activity was recognized in India from January to April by several authors (Jerdon 1874, Baldwin 1876, Brander 1923, Stockley 1928), and Schaller (1967) spoke of a second peak from mid-August to mid-October. In the Serengeti gazelles, a peak reproduction period begins with the end of the small rains (usually mid-December) and is continued with some fluctuations until the end of the long rains in May. Another peak seems to be in September. However, there can be considerable variations due to differences in local conditions (e.g., onset, length, and intensity of the rains, etc.) in different years. Thus, one can only say that there is a low in reproductive behavior from the beginning of June to roughly mid-August in all these species.

As stated above, the peaks and lows of territoriality are somewhat linked to the peaks and lows of reproduction, but they do not precisely coincide. Increases in territoriality can precede peaks in reproduction, and, on the other hand, males may still be holding their territories after a peak in reproduction is over. It can even happen that while some of the males holding territories during a peak in reproduction may abandon them after the end of this period, other, previously non-territorial males may take advantage of the now vacant space and establish territories within it. Due to this process, the number of territorial males can remain constant or even somewhat increase after the end of a reproductive peak.

Figure 18: Solitary territorial males. (a) Thomson's gazelle. (b) Grant's gazelle. (Photos: F.R. Walther—Serengeti National Park, Tanzania.) (c) Blackbuck resting on top of a dung pile. (Photo: E.C. Mungall—Greenwood Valley Ranch, Texas.)

Territoriality and Population Density

As long as a male holds a territory, he does not participate in the movements and migrations of the herds. Therefore, it can happen that most of the females and non-territorial males leave a given area, but many of the territorial males remain. Then, they may make up a considerable portion of the population that is left. On the other hand, when the moving herds invade an area, the portion of territorial males can be relatively low, even during a peak in territoriality. Thus, the proportion of territorial individuals within a population, or even only among the adult males, may vary greatly in a limited area during the course of the year. When trying to determine the percentage of territorial individuals in a given area, one must be aware that it is not a constant figure in free-ranging animals, particularly when seasonal movements take place. To illustrate these statements, the situation of Thomson's gazelle in the observation area in the Togoro Plains in 1965 and 1966 may be presented (Table 2).

The total number of Thomson's gazelle ranged from 155 (or even only 11 when the extreme case of August 6, 1966 was included, i.e., the day after the whole area had burnt) to 2,994. The number of (territorial and non-territorial) adult males ranged from 53 (or five on August 6, 1966) to 281, and the numbers of territorial males from eight (or three on August 6, 1966) to 129. The territorial and non-territorial adult males composed an average of 25.2% of the population with a minimum of 9.1% and a maximum of 42% (45.5% on August 6, 1966). The average proportion of territorial males in the total population was 9.1% with a minimum of 2.5% and a maximum of 22.8% (27.3% on August 6, 1966). Territorial males made up an average of 36.9% of the adult male population with a minimum of 12.7% and a maximum of 77.6%.

In some aspects, the extreme case of August 6, 1966, was particularly instructive. After the intensive burning the day before, the feeding conditions were extremely poor in Togoro. Consequently only a very few tommy were found there. However, with respect to the very small total, the proportions of adult males (45.5%) as well as of territorial males (27.3%) were the greatest ever recorded in this area. Also, the percentage of territorial individuals within the adult males (60%) was well above the average (36.9%), although it was somewhat exceeded by the corresponding proportions during the last months of this year which presented another special case.

Likewise, although not as extreme, the months of July and August in both years showed small population sizes (ranging between 155 and 276 animals) due to unfavorable environmental conditions (dry season and the grass grown high in large parts of Togoro) and the absence of migratory herds passing through the area. The two greatest proportions of adult males in the population (40% and 42%) occurred during these four months, and the proportions in the two other months (29.3% and 34.2%) were only reached or slightly exceeded in seven of the 18 remaining months. Also, the portion of territorial males was great (12.5% to 15.2%) in the population in three out of the four months under discussion and was reached or somewhat exceeded only in October through December 1966.

On the other hand, the greatest population densities of more than 1,000 to almost 3,000 animals were reached in February, March, November, and De-

Table 2: Thomson's Gazelle Population and Territorial Males in the Togoro Plains Observation Area (Serengeti National Park) During 1965 and 1966

Year	Month	Day	Total Population	Adult ♂♂ (terr. + non-terr.)	Terr. ♂♂	Proportions of Adult ♂♂ in Total Population	Proportions of Terr. ♂♂ in Total Population	Proportions of Terr. ♂♂ Among Adult ♂♂
1965	Jan.	22	729	128	37	17.6%	5.1%	28.9%
	Feb.	27	1,118	123	55	11.0%	4.9%	44.7%
	Mar.	10	2,145	281	53	13.1%	2.5%	18.9%
	April	20	948	184	76	19.4%	8.0%	41.3%
	May	19	869	185	72	21.3%	8.3%	38.9%
	June	7	332	116	33	34.9%	9.9%	28.5%
	July	16	250	100	37	40.0%	14.8%	37.0%
	Aug.	17	208	61	26	29.3%	12.5%	42.6%
	Sept.	16	809	253	32	31.3%	4.0%	12.7%
	Oct.	14	467	90	38	19.3%	8.1%	42.2%
	Nov.	8	2,994	272	97	9.1%	3.2%	35.6%
	Dec.	3	1,938	275	129	14.2%	6.7%	46.9%
1966	Jan.	5	868	166	61	19.1%	7.0%	36.8%
	Feb.—No count							
	March—No count							
	April	10	512	166	36	32.4%	7.0%	21.7%
	May	4	543	197	32	36.3%	5.9%	16.4%
	June	5	558	165	31	29.6%	5.6%	18.8%
	July	8	276	116	42	42.0%	15.2%	36.2%
	Aug.	(6)*	(11)	(5)	(3)	(45.5%)	(27.3%)	(60.0%)
		11	155	53	8	34.2%	5.2%	15.1%
	Sept.	8	1,669	222	74	13.3%	4.4%	33.3%
	Oct.	19	729	163	122	22.4%	16.7%	74.9%
	Nov.	4	501	147	114	29.3%	22.8%	77.6%
	Dec.	5	404	142	89	35.2%	22.0%	62.7%
Average**			864.6	163.9	58.8	25.2%	9.1%	36.9%

*One day after a burn in the entire area

**Excluding the count on August 6, 1966

Abbreviations: Terr. = territorial; non-terr. = non-territorial cember 1965, and in September 1966, due to extensive rainfall and the availability of short, green grass. In all these months with particularly favorable environmental conditions and invasions of large herds, the proportions of adult males (ranging between 9.1% and 13.1%) as well as those of territorial males (between 2.5% and 4.4% of the population) reached their lowest values.

In order to interpret the figures from August to December 1966, one must know that there were a few local rain showers in Togoro in September resulting in the growth of fresh, short grass in this area after the total burning in August. During the following months, however, there was no rainfall in Togoro, and the country dried up again. There were good rains in other parts of Serengeti during this time. Consequently, there was an invasion of gazelles into Togoro in September, but later more and more gazelles left this area, and the Togoro population dropped continuously from 1,669 in September to 404 in December. With the decline of the population size, the proportion of adult males went up from 13.3% to 35.2%, and that of territorial males within the total population from 4.4% to 22% and slightly more. The December population (404 animals) was comprised of 262 females plus immatures and juveniles of both sexes and 142 territorial and non-territorial adult males. When the enlarged September population with 1,447 females and young animals and 222 adult males is taken as a reference figure, it appears that the proportion of females and young had declined by 72%, but that of adult males by only 36%, and within these adult males, 62.7% to 77.6% were territorial during the period of October through December.

These data suggest the following generalities, at least under the Serengeti conditions: (a) The proportion of adult males in general and that of territorial males in particular are negatively correlated to the size of a population within a definite, limited area, i.e., the smaller the population, the greater the proportion of adult and territorial males, and vice versa. (b) The females and young animals often respond more quickly to unfavorable changes in the environment and many of them leave an increasingly unfavorable area earlier than adult males and particularly territorial males.

In contrast to the proportions, the absolute numbers of territorial males are positively correlated to population size, i.e., the greater the population, the greater are the numbers of territorial males apparently up to a maximum capacity given by the local spatial conditions and limitations. (In Togoro, this maximum seemed to be 120 to 130 territorial tommy males in those years.) As can be taken from Table 2, however, this is only a rule of thumb. In an area with a changing population density, the peaks in the numbers of territorial individuals lag behind the peaks of population size. For example, on February 27, 1965, the total population was 1,118 animals with 55 territorial individuals in Togoro. On March 10, 1965, the total had increased to 2,145 animals, i.e., almost double the former; however, the number of territorial males was only 53, i.e., about the same as in February. An increase to 76 territorial males was found in the next month, on April 20, 1965, when the total population was already dropping (948 animals). To give another example, on October 14, 1965, the total was 467 animals with 38 territorial individuals. On November 8, 1965, a strong invasion of migrating herds brought the total number of tommy up to 2,994, and with it, the number of territorial males increased to 97 in Togoro. However, on December 3, 1965, the number of territorial males had risen to 129, whereas the total population had declined to 1,938. A similar situation was found in September/October 1966, and also some other data presented in Table 2 (e.g., June to August 1965, or June/July 1966) suggested that the decrease in the absolute numbers of territorial individuals often only hesitantly followed the decrease in the total number of animals. In short, the increases as well as the decreases in absolute numbers of territorial males within a given area are positively correlated to the increases or decreases in population size, but it is not an immediate correlation since the increases and decreases in territorial individuals lag somewhat behind those in the total numbers of animals.

In Grant's gazelle, essentially the same principles were found, although, of course, in smaller dimensions due to the smaller population and the larger size of Grant's territories, and thus, the smaller numbers of territorial individuals in limited areas. For example, in Togoro during the same period as discussed above for Thomson's gazelle, the total numbers of Grant's gazelle ranged from 10 to 300 with an average of 92.9, and the numbers of territorial bucks ranged from one to nine with an average of 4.5. The average proportion of territorial males from the total population was 5.9% and the maximum proportion was 10.1%, i.e., this maximum approximated the average in Thomson's gazelle (9.1%). On the other hand, the average proportion of territorial individuals within the adult males was 39.2% in Togoro, and thus, approximately the same as in Thomson's gazelle. During unfavorable periods, this figure even became 100%, i.e., all the Grant's gazelle males who remained in the area were territorial. In open plains areas of a comparable size, such as the Ngare Nanyuki area or the area around the Simba Kopjes, but with larger herds of Grant's gazelle, the average proportion of territorial males from the total of the population was even smaller (2.1%) and so was the average proportion of territorial individuals among the adult males (9.5%).

TERRITORIALITY VERSUS GREGARIOUSNESS AND MIGRATORY BEHAVIOR

The above discussion indicates certain, mainly negative, correlations of territoriality with gregariousness and migratory behavior. Basically, blackbuck and gazelles are gregarious animals, but territorial males leave the herds and isolate themselves from the others. In species such as Grant's and Thomson's gazelles in Serengeti, the majority of the animals perform seasonal migrations. This behavior, in a sense, is incompatible with territorial behavior in that the owner of a territory either has to abandon his territory in order to move with the herds or he has to remain in his place and cannot participate in the migration. Linked to the last point is the relative proportional increase of territorial individuals in a given area when the majority of the animals have left. On the other hand, the absolute numbers of territorial individuals usually increase when migratory herds have arrived in a given area, i.e., a number of the migratory males leave the herds and establish territories when they arrive in a new area. This is the positive correlation between migratory behavior and territoriality.

These are the major points discussed thus far which indicate the dynamic interactions between territoriality and migration. However, this interaction runs deeper. Besides the general internal readiness to migrate and the environmental conditions which release the migration, and besides the local traditions, the landmarks and other factors which orient it, there are two social factors in these animals which keep the migration going. First there is the "pulling" effect of the preceding animal(s), i.e., the tendency to follow a moving conspecific. Secondly, the "pushing" effect of the animal(s) following behind (Walther 1978a,b) propels the migration. Pushing particularly occurs when animals start moving after a resting or a grazing pause, or when some of the marching animals slow down, come to a halt, or deviate from the mutual course. The pushing is brought about by aggressive displays (threats, dominance displays) and even by fighting and chasing. In Thomson's gazelle, pushing is also effected by certain "courtship" displays (neck-stretch, nose-up) which can substitute for threat displays in this situation when the sender is a male and the recipient is a female (Walther 1978b). The non-territorial adult males in the mixed herds are the most active animals in such pushing actions. They may be said to be the "motors of migration." When some of these males become territorial, they no longer participate in the pushing actions. On the contrary, they now try to prevent moving females from leaving their territories, and they chase away migratory males passing through the territories. If they are very successful, they may temporarily separate the non-territorial males from the females. In these ways, migratory herds lose their "motors." One may say that the territorial males are the "brakes of the migration." How successful they are in this "braking action" depends—besides the affects of local conditions (e.g., dryness of the area), numbers and sizes of migratory herds, and strength of their migratory mood—on the numbers of territorial bucks relative to those of non-territorial adult males in the mixed herds. When there are only a few territorial males in a given area and large migratory herds with many non-territorial adult males pass through, a territorial male may exhaust himself in an outburst of fruitless herding and chasing actions. Finally, he gives up, stands there or even beds down and lets the waves of migration pass through his territory. Sooner or later, he usually abandons his territory and moves away altogether with the migrating herds. However, when a good number of males have established a territorial mosaic in a given area and when relatively few or small herds invade, the actions of the territorial males can be very effective and can bring the migration to a standstill. When the migratory herds have reached areas with favorable feeding conditions and now move ahead within these, more and more adult males leave the herds and become territorial, and their retarding effect upon the migration becomes stronger and stronger. In such a situation, territoriality is the direct antagonist of migratory behavior.

Another relationship between territoriality and migration becomes evident in special cases. As mentioned above, it can happen that the majority of the gazelles have left a given area but some of the territorial males remain. Later, local conditions may become worse and worse, but the bucks may still hold their territories. When a migrating herd subsequently enters this area and moves through the territories, these males may abandon their territories and migrate with their conspecifics leaving all the territories empty after the herd has passed through.

Similar to the described exhaustion of the male's "territorial energies" by large migratory herds, a territorial male outside the migration season also may consume his energies and quickly abandon his territory when it is located at a site where huge numbers of conspecifics continually pass through on their way to a definite locality. This can happen, for instance, when the territory is located near a highly frequented waterhole.

Furthermore, one may sometimes speak of a certain "softening" of territorial behavior as a consequence of great population density or large herd size. For example, a territorial Grant's gazelle buck can keep a relatively stable harem and can reliably exclude non-territorial males from his territory only as long as he deals with relatively small groups of conspecifics. When mixed herds invading his territory approach numbers of one hundred or more animals, he "restricts himself" to dominating the other males as long as they are within his territory and to herding a pseudo-harem of females for a couple of hours.

Such negative correlations of gregariousness and migratory behavior with territorial behavior, however, do not exclude the existence of territoriality in a migratory and/or gregarious species. The gazelles and their relatives offer impressive testimony of this fact. It is perhaps not out of place to stress this point since an opinion has been expressed in the literature that gregarious ungulate species would not be territorial (Lorenz 1963). This view certainly is mistaken. Territoriality and gregariousness can co-exist, and one may say that territoriality serves to overcome certain negative aspects of crowding in highly gregarious animals.

This last statement is not merely a theoretical consideration. Corresponding events can be directly observed in Thomson's gazelle, Grant's gazelle, mountain gazelle and Indian blackbuck. For example, a female in a mixed and large herd of Thomson's gazelle may sometimes apparently come into estrus even during a migration period. Then several males, up to 15 have been observed in this species, are immediately after this female and court and drive her incessantly. As will be discussed later in detail (p. 142), an Antilopinae female's mating behavior is nothing but a moderately ritualized withdrawal. Consequently, her escape reaction can be "raised above the thresholds" very easily, which may sometimes happen even when only one (territorial) male is courting her. Exceeding the "threshold" is certain when several males are after her, and she flees as fast as she can. All the males follow at a full gallop, but they do not try to pass her and to block her path as a territorial male would do within his territory (see p. 137). Thus, unobstructed, the female runs ahead with increasing speed. This situation is anything but favorable for courting and mating. Moreover, the chasing males almost regularly start fighting with each other which, of course, increases the female's chances of escape, and in a large herd, such a female very skillfully uses the other females as a cover and hides among them. Theoretically, copulation is not absolutely impossible in such cases; however, it has never been witnessed in any of the species under discussion by any observer during the years of study. On the other hand, when a male has left the herd, has established a territory and has kept it clear of sexual competitors, and when he now herds a relatively small group of females in his territory, his chances for successful courting and mating are incomparably greater.

4

Shape, Size and Density of Territories; Territorial Periods

GENERAL ASPECTS

Shape of Territories

The shapes of Antilopinae territories vary considerably even within the same species. Environmental factors such as landmarks and differences in the suitability of the occupied ground, social factors such as the shape of neighboring territories, as well as "historical" aspects (e.g., a male has formerly held a territory in the same area) may play a role and account for the great variations with respect to shape. Consequently, only a few general principles can be given. If generalized in terms of geometric figures, some of the territories approximate a circle or a polygon; however, ovals and rectangles are more common, i.e., many territories are more extended in one dimension than another. Furthermore, more or less pear-shaped territories are relatively frequent, i.e., they are broader toward one end than toward the other. Finally, there are always territories which do not fit any of these categories, and about which one cannot say more than that they are of irregular shape (see Figure 19).

Size of Territories; Upper and Lower Size Limits

With respect to territory size, a general rule says that big animals have large territories and small animals have small territories (Hediger 1949). This generally is true when one thinks of such differences in body size as, for example, that of a hamster and a rhinoceros. However, when the difference is not so enormous, the above rule often fails or, at least, has quite a number of exceptions. Of course, a Grant's gazelle is much bigger than a Thomson's gazelle, and also its territories are considerably larger than those of tommy. On the other hand, a mountain gazelle is not much bigger than a tommy, and, certainly, it is much smaller than a Grant's gazelle. However, the territories of mountain gazelle even exceed the Grant's territories in size.

Another question is whether the density of territories may influence their size. Due to technical difficulties (e.g., in some of the species investigated the conditions for such a comparison simply were not present in the areas during the observation periods; in other cases, promising conditions were available but the investigation would have taken an exorbitant amount of time, etc.), none of us has taken precise records concerning this problem. However, some general aspects can be given.

Primarily, the size of the territories seems to be a species-specific matter. For example, the largest tommy territories observed were still somewhat below the size of the smallest territories in Grant's gazelle. Thus, one may safely say that Grant's territories are larger than tommy territories regardless of their density. Within this species-specific realm, territorial density can have influence on size. In an area with a few territories, the size of the territories may be larger than in the same or a comparable area with a dense territorial mosaic. However, the extent to which a territory can be enlarged even when few or no neighboring territories are around, or the extent to which it can be compressed under dense conditions, is apparently limited.

Schenkel (1966) proposed the idea that a territorial animal must be able to keep the occupied area under constant visual control. This requirement does not appear true in Antilopinae. For example, Grant's gazelle can successfully establish territories in scattered woodland, or a territory may include a small hill or a depression. Thus, the owner cannot keep his entire territory under constant visual surveillance. However, one may accept a modification of Schenkel's suggestion in that a territorial buck cannot keep an area of a larger size than that in which he can fulfill the activities necessary for maintaining his territorial status. This requirement puts a limit on the size of a territory, and this limit varies with the species-specific territorial activities. For example, of the Antilopinae species under discussion, Thomson's gazelle show the most intensive marking with preorbital gland secretion. In a well-established tommy territory, there may easily be 100 or more secretion marks, and they have to be continually renewed. This intensive marking by preorbital gland secretion obviously belongs to the territorial activities of a tommy male, and it seems to be essential for his territorial status. However, the limits in producing preorbital gland secretion set limits for the area which can be intensively marked this way. It is probably more than coincidence that this species, in which preorbital gland marking plays such a great role, has smaller territories than most of the other species under discussion.

Corresponding aspects are valid for all the other territorial activities such as establishing dung piles, herding females, chasing away or, at least, dominating non-territorial male intruders, challenging or stopping territorial neighbors at the boundary, etc. In short, when a territory is enlarged beyond a certain optimum size, the owner cannot fulfill his territorial "duties" in the entire area. Thus, he either has to give up parts, or he exhausts himself in a short time and ceases his territorial status completely.

In species such as Grant's and mountain gazelles with their large territories, the animals sometimes apply a "trick" to keep an area larger than that in which they can accomplish their territorial activities. They temporarily use "core areas," i.e., the owner more or less restricts his activities to a section of his territory for some time, and he later shifts his activities to another part of the territory. Understandably, this "strategy" also has its limits.

We emphasize that the above are not merely theoretical considerations. Corresponding events can actually be observed. For example, in Thomson's gazelle with their dense territorial mosaics, it was noted that many of the territorial males left a given area during the dry season, but some of them remained. In this situation, some of the remaining males extended their territories into the recently vacated space. Those bucks who only moderately extended their territories, kept them. However, those who enlarged their territories to double or even more of their former size, regularly abandoned the area after a few days. During these last days they gave the impression of not being sure of their territorial status anymore (p. 184).

In short, the necessity to fulfill territorial activities sets the upper limit of territory size within the species-specific realm. When seeking factors which set lower limits, one could think of the animal's feeding requirements; it could be postulated that any territory must be at least large enough to guarantee the owner a sufficient food supply. As logical as this argument may first appear, it does not hold true for ungulates. For example, the territories of Uganda kob *(Kobus (Adenota) kob thomasi)* in the "arena" situation described by Buechner (1961), are so small that the owners regularly leave them for a few hours each day in order to feed. Also, we will have to discuss comparable cases in Antilopinae later (p. 187). Thus, in ungulates, a territory can be smaller than what would be needed to fulfill the owner's feeding requirements. More important with regard to minimum territory size are courtship activities. Obviously, territoriality of "territorial/gregarious" species (Estes 1974) has connections with reproductive behavior. For reproductive success, the male must be able first to herd females within his territory, and secondly, to elicit a "mating march" in most of these species. When the territory does not have sufficient size to allow a male to court and drive a female according to the species-specific mating ritual without leaving the territory, or when the female has to walk only a few steps to pass the boundary and to escape from the male's herding efforts, the territory is definitely too small. According to this reasoning, one may say that the fulfillment of the owner's activities may usually also set the limits for minimum territory size.

Except for a very special situation of blackbuck territories at Velavadar National Park in India (p. 65), territories considerably smaller than the average were rare in our studies, and apparently were due to rather unusual situations. Two observations in Thomson's gazelle may illustrate this point. In the one case, a male had been territorial for several months. Later he moved off together with the migratory herds, and returned to the same area after about four months absence. Meanwhile, two other males had established territories there which included almost the entire territory of the former male. Upon returning, the male established a "mini-territory" of hardly 30 m in diameter in an unoccupied corner between the two territories. After three days, he started conquering back his former holdings from this base. The second case involved an apparently senile male who had established a territory within a mosaic of other territories and who kept his place after the majority of the gazelles—including his neighbors—had left the area at the beginning of the dry season.

With the onset of the rains, the herds returned to this area, and many males established new territories there. In the boundary encounters with his new neighbors, the old male could not keep up. In the course of three weeks, two of his neighbors gradually enlarged their territories into the space occupied by the senile male, "crushing" his territory between them. When it was down to less than 3,000 m^2, about a quarter of its original size, the old buck abandoned it. These two cases were exceptional. More commonly one may observe that a tommy buck holding his territory during the dry season after his neighbors have left, expands his territory into the newly vacant space, but gives up the additional space and retains only his original territory when the herds return and many males establish new territories around him.

From these and similar situations, one may conclude that the range between maximum and minimum size of a territory is not too great within a species and that a territory cannot be compressed very much below average size in most of the cases. Enlarging the territory when the area around it is vacant, is quite usual, but this process also has relatively narrow limits.

Density of Territories

The relatively small deviations from an optimum species-specific average size determine the maximum densities of territories in a given area. Provided that environmental conditions are equally suitable throughout the area, one could divide the space available by the species-specific average territory size and get a fair approximation of the maximum number of territories that can be established there. The actual number of territories is likely to approach the theoretically determined maximum capacity under favorable environmental conditions and with an adequate population. For example, blackbuck territories on Texas ranches have an average size of 4 ha. In a large pasture, the entire area used by blackbuck (including females and non-territorial males) covered 66 ha, and there were 12 territories in it. This number compares quite well with the theoretical maximum of 16 territories.

When the maximum capacity or the optimum (see below) is reached in a given area and there are still non-territorial males present, the latter do not normally try to conquer occupied territories or parts of them. (An exception is given when one of these males has been territorial in this area previously.) If parts of the area are less favorable for territories and presently are unoccupied, some of the surplus males may try to establish territories there. Usually such "experimental" territories are abandoned after a few days or even a few hours. Most of these males remain in the vicinity as non-territorial bachelors, or leave and move to other areas. If one of the territories becomes vacant, a bachelor male may occupy it, but it is by no means rare that an abandoned territory remains empty for days and even weeks in spite of the presence of non-territorial adult males in the vicinity.

Length of Territorial Periods; Re-Occupation of Territories

The length of the territorial periods varies among species and even considerably within each species due to local, environmental, and social situations, and probably also due to the individual condition of the territory owner (age of the animal, its general physical condition, strength of its attachment to the area based on such criteria as previous territorial stays there, familiarity with the neighbors, etc.). Among the species investigated in our studies, mountain gazelle and Thomson's gazelle seem to represent the two extremes.

In mountain gazelle, only three out of 13 individually known males were territorial for periods less than one year (one month, six months, nine months). The others were observed for more than 12 months in their territories and did not show any signs of leaving at the end of the study. In Thomson's gazelle, only two of the observed males stayed uninterruptedly in their territories for approximately one year, and the average of 52 recorded territorial periods was about two and a half months. It is likely that this average figure is even biased toward longer territorial periods due to a human factor. Tommy males quite frequently try to establish territories under unfavorable conditions and abandon them after a few days or even a few hours. Naturally, such cases are much harder to recognize than long term stays in a territory, and thus, the number of short term territories in the sample observed may easily be too small a proportion.

Seven out of 44 tommy males returned to their former territories or the immediate vicinity after periods of absence. One of them even returned twice. Two of the absences lasted for only one to two weeks, while the others ranged from one month up to 10 months. Again, a human factor cannot be completely excluded in that possibly some more males returned to former territories but were not re-identified by the observer, particularly when they had held their former territories for only very short periods.

As stated above, mountain gazelle and tommy represent two extremes among the species investigated. The other species are intermediate, i.e., they are not as permanently territorial as mountain gazelle, but, on the average, their territorial periods are longer than those in Thomson's gazelle. Returns to formerly held territories are at least as frequent as in tommy. Re-occupation of a territory by the same male must be distinguished from occupation of the same territory by another male. The latter was particularly evident in blackbuck studies on Texas ranches where several times another male took over after the former owner had left his territory. Interestingly enough, the new owners usually kept precisely the boundaries of the pre-existing territories. Also, some blackbuck males at least tried to establish new territories in other areas not long after they had left or lost their former holdings.

SIZE AND DENSITY OF TERRITORIES AND TERRITORIAL PERIODS IN SINGLE SPECIES

Thomson's Gazelle

The following data on Thomson's gazelle (Table 3) were mainly gathered in the Togoro Plains in 1965. The area was mapped at the beginning of that year, and a map was prepared on the scale 1:1,000 cm. In a section of this area, the boundaries of 13 neighboring territories (Figure 19) were determined by direct observation of agonistic encounters between neighbors, beginnings and endings of chasing actions toward trespassing bachelors, onsets and endings of herding activity toward females, and the positions of dung piles, in March and April

1965. All these males were individually known, and all except one had been in their territories for at least one month. Three of them abandoned their territories before the end of April, all the others were permanently in their territories throughout the two months under discussion (and longer). In order to figure the sizes of 13 territories, their boundaries were copied on squared millimeter paper, and the squares (1 cm^2=100 m^2) were counted.

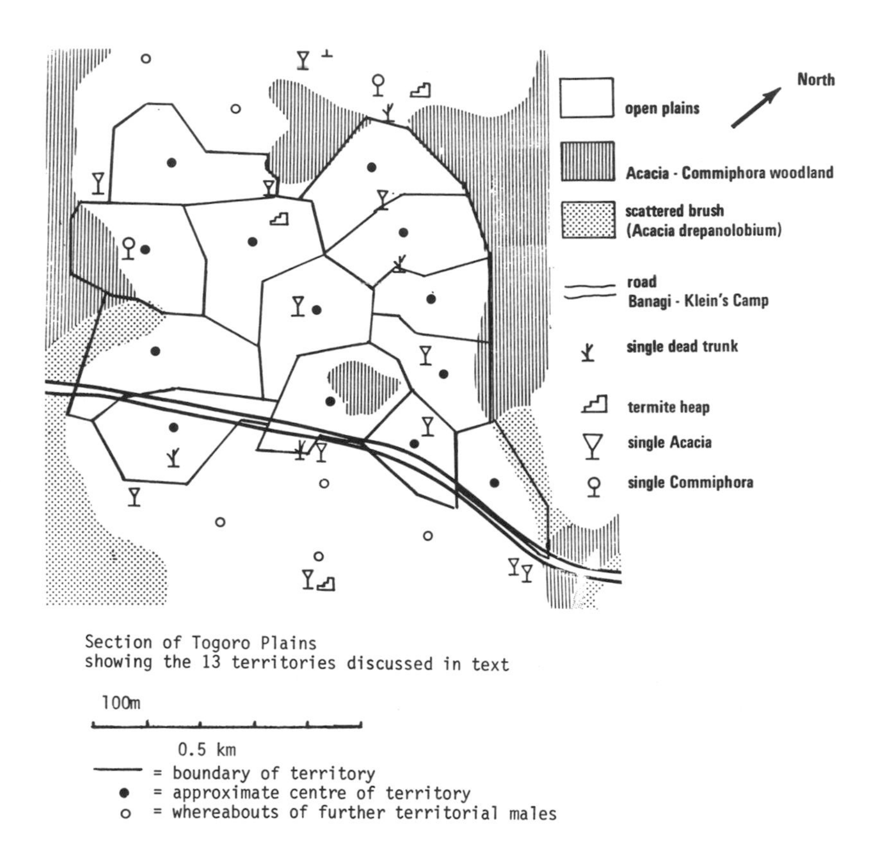

Figure 19: Map of a section of the study site in Togoro (Serengeti National Park, Tanzania) with territorial mosaic of Thomson's gazelle.

The investigation on lengths of territorial periods is based on the observation of 52 territorial periods of 44 males (some of them became territorial a second time after a period of absence) in Togoro in 1965 and 1966, most of them between February and November 1965 (Table 3). The males were individually identified mainly by using the variations in the black nose patches and their white forehead patches in combination (see p. 17 and Walther 1973b). Each male was checked at least once, but usually two to three times per week. Four of these males had been territorial before the observation began; however, their behavior strongly indicated that they were in the very beginning stages of territoriality (see p. 182). Since all four males remained in their territories for more than four months, their territorial time before the beginning of the observation (probably only a very few days if not hours) was considered to be insignificant, and their territorial periods were counted from the first day of observation. All the other males were observed throughout their entire territorial periods, i.e., from the very beginning to the very end.

Table 3: Territory Sizes and Territorial Periods* in the Antilopinae Species Investigated

	Territory Sizes (ha)	Territorial Periods (months)	Remarks
Indian blackbuck	0.3–20 m.c.: 2–9	<1->17 m.c.: ½–8	N territories: 39 N territorial periods: 20 (14 of them incomplete) including 4 re-occupations
Thomson's gazelle	0.3–10 m.c.: 1–5	<1->11 m.c.: ½–5	N territories: 17 N territorial periods: 52 (5 of them incomplete) including 8 re-occupations
Grant's gazelle	15–60	4->11 m.c.: 4–8	N territories: 4 N territorial periods: 10 (3 of them incomplete) including 3 re-occupations
Mountain gazelle	100–220 core areas: 25–50	>1->20 m.c.: probably several years	N territories: 7 N territorial periods: 13 (all of them incomplete)
Springbok	27–70	4->12	After Mason (1976) N territories: 3 Territorial periods are estimations
Gerenuk	130–340	16–41	After Leuthold (1978) N territories: 10 N territorial periods: 5

Abbreviations: N = numbers of territories measured or territorial periods recorded
m.c. = most commonly (more than 60% of the observations)

*Territorial period means uninterrupted stay in the same territory. Re-occupation of the same territory by the same male after an absence of weeks or months, was counted as another independent territorial period. Brief absences of a few hours per day were not considered to be interruptions of the territorial period.

As previously mentioned (p. 56), a territory once was "crushed" between neighboring territories, and the male abandoned it when it had dwindled down to about 0.3 ha. Thus, this territory obviously represented the minimum size; one may even say that it had become too small. The next smallest tommy territory observed (in an area other than Togoro) was about 0.8 ha. It was well-established, the owner showed all the normal activities of a territorial male, and he kept it for more than two months. The two largest tommy territories which were ever seen (also in areas other than Togoro) approximated 10 ha. However, the owners of these large territories clearly showed signs of declining territoriality (p. 184). Thus, these territories obviously represented the maximum; one may even say they were too large.

The sizes of the 13 tommy territories investigated in the Togoro Plains in March and April 1965, ranged from 1.5 ha to slightly more than 5 ha, with an average of about 3 ha (about 7.5 acres). At maximum (December 3, 1965), 129 territorial males were found in the entire observation area. Thus, about 390 ha (3 ha per territory) were occupied by territorial males at that time. Excluding the densely vegetated areas where tommy males usually do not establish territories (p. 70), the open shortgrass plains in the entire observation area approximated 400 ha. These figures suggest that the males made use of all the space suitable and available in that area at that time. However, such a high number of territorial males and with it, the use of the total space available, were only reached in two out of 22 months (Table 2). Thus, this density of the territories seems to represent the maximum, but probably not the optimum. Theoretically only 176 ha (i.e., less than 50% of the suitable area) were occupied when one considers the average number (58.8—see Table 1) of territorial males in this area (again with an average size of 3 ha per territory). This figure probably comes much closer to the optimum.

The 52 territorial periods recorded (Table 3) ranged from not quite one full day up to a few days more than 11 months. 11.5% of them lasted less than one week, 17.3% ranged from one to four weeks, 44.2% lasted between one and three months, 15.4% from three to six months, and 11.5% from seven to eleven months. These figures indicate that when a tommy buck has established a "good" territory, which usually takes him at least one week (see p. 85), there is a good probability that he may keep it for several weeks or months, most commonly for a period between two weeks and five months, and occasionally he may even stay without any interruption in the same territory up to about one full year. On the other hand, one may not forget that, as already discussed, these proportions probably are somewhat biased toward the longer periods since very short term territories can easily escape the attention of an observer.

Grant's Gazelle

In the same observation area (Togoro Plains) and during the same period (1965-66) as described for Thomson's gazelle, the number of territorial males of Grant's gazelle ranged from one to nine, with an average of 4.5. Thus, there were considerably fewer territories than in Thomson's gazelle. The smaller number of territories may partially be due to the generally smaller population of Grant's gazelle in Serengeti (p. 28), but the larger size of the Grant's territories seems to be the most important reason. Four Grant's territories were mapped and measured in Togoro (Figure 20). The largest of them was situated almost precisely in the same location as the 13 tommy territories described above and was even somewhat larger (about 60 ha) than all of them together. The next largest territory was about 45 ha. The two other territories were considerably smaller with about 20 and 15 ha. Territories of Grant's bucks observed later in the central Serengeti Plains, were not mapped or measured, but the sizes of four of them were estimated. They were in about the same range as the two large territories in Togoro. Generally, it may be safe to say that Grant's territories range from a size of about 15 ha to 60 ha, with the truly "good" territories near the upper end of this range (Table 3).

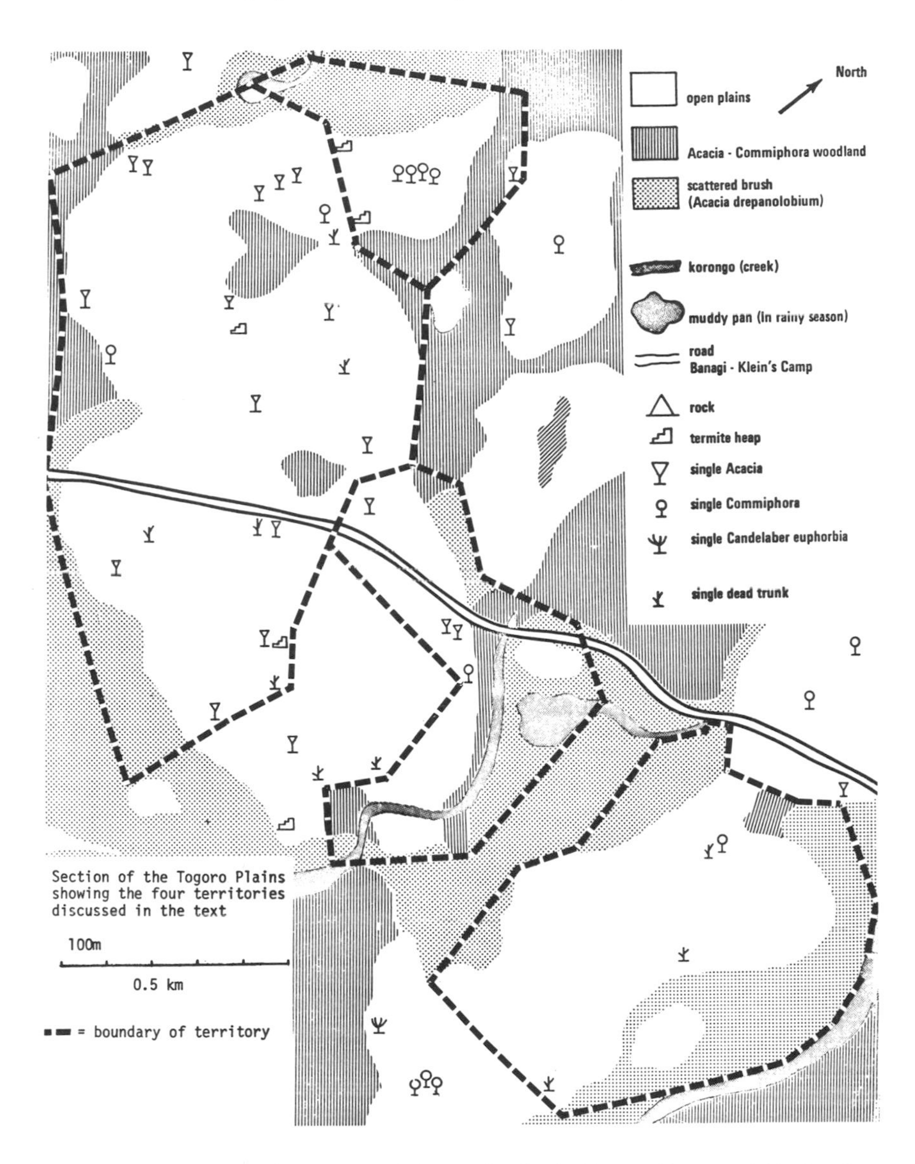

Figure 20: Map of a section of the study site in Togoro (Serengeti National Park, Tanzania) with four neighboring territories of Grant's gazelle.

The two small territories in Togoro clearly belonged to less successful territory holders. The smallest territory (15 ha) belonged to a young-adult buck who probably had become territorial for the first time in his life. He was alone for a long period, and later his territory was only occasionally frequented by females. He never established a more permanent harem as did the males with the larger territories in Togoro. The other small territory (20 ha) belonged to a fully adult buck, but apparently his territory was poorly located. About half of it was covered with bush and forest whereas the "good" territories were either completely in the open plains or only framed by woodland. A somewhat stable harem was with this male only during a comparatively short period of his total territorial time.

The young-adult male in Togoro mentioned above, became territorial in January 1966 and kept his territory until the beginning of August, i.e., for about seven months. The adult buck with the poorly located territory, was seen there from May through October 1965; however, he had probably been in his territory before May. He came back to his territory in January 1966 and was present until September 1966, i.e., a period of nine months. One of the "good" territorial males was not individually identified before May 1965, although he probably had already been there for some time. He was in his territory from May to August 1965. Then he disappeared for about one month, but he was back in mid-September in the same territory and kept it uninterruptedly until early August 1966, i.e., for about 11 months. The other very successful territorial buck (Table 1) in Togoro was observed from the beginning of January 1965; however, he also may have established his territory some time before. He kept his territory into the first days of September 1965, i.e., for at least eight months. In the following three months, he was absent most of the time, but he sporadically reappeared on his territorial grounds for a few days during this period. He stood firmly in his territory again from mid-December 1965 through the first days of August 1966, i.e., another eight months. In addition, the territorial periods of three males were recorded near Ngare Nanyuki. Two of them were in their territories from mid-May through mid-September 1966, i.e., for four months. The third buck was territorial from mid-May through mid-November 1966, i.e., for six months; however, he had somewhat shifted his territory from its original location during the last one and a half months.

On the whole, the maximum territorial periods of uninterrupted stay in the same territory were not found to be longer than those in Thomson's gazelle (not quite one year). However, the observed territorial periods in Grant's gazelle ranged from four to 11 months, with relatively many of them between six and nine months. Thus, the territorial periods of Grant's gazelle seem to average somewhat longer than those of Thomson's gazelle.

Mountain Gazelle

Because of their large size, the territories of mountain gazelle in the northern Negev were not measured. Size estimates were made by plotting the boundaries on a map of known scale (Figure 21a). The size estimates of seven territories varied from approximately 100 ha to 220 ha with an average of about 150 ha (Table 3). Most of the male's daily activity was confined to a smaller area of only 25 ha to 50 ha within his territory (core area). The territories appeared to be somewhat smaller and closer together in the rough terrain of Ramat Yissakhar where the population also was denser than in the Negev. However, complete territory boundaries could not be determined here because of disturbance to the population early in the study.

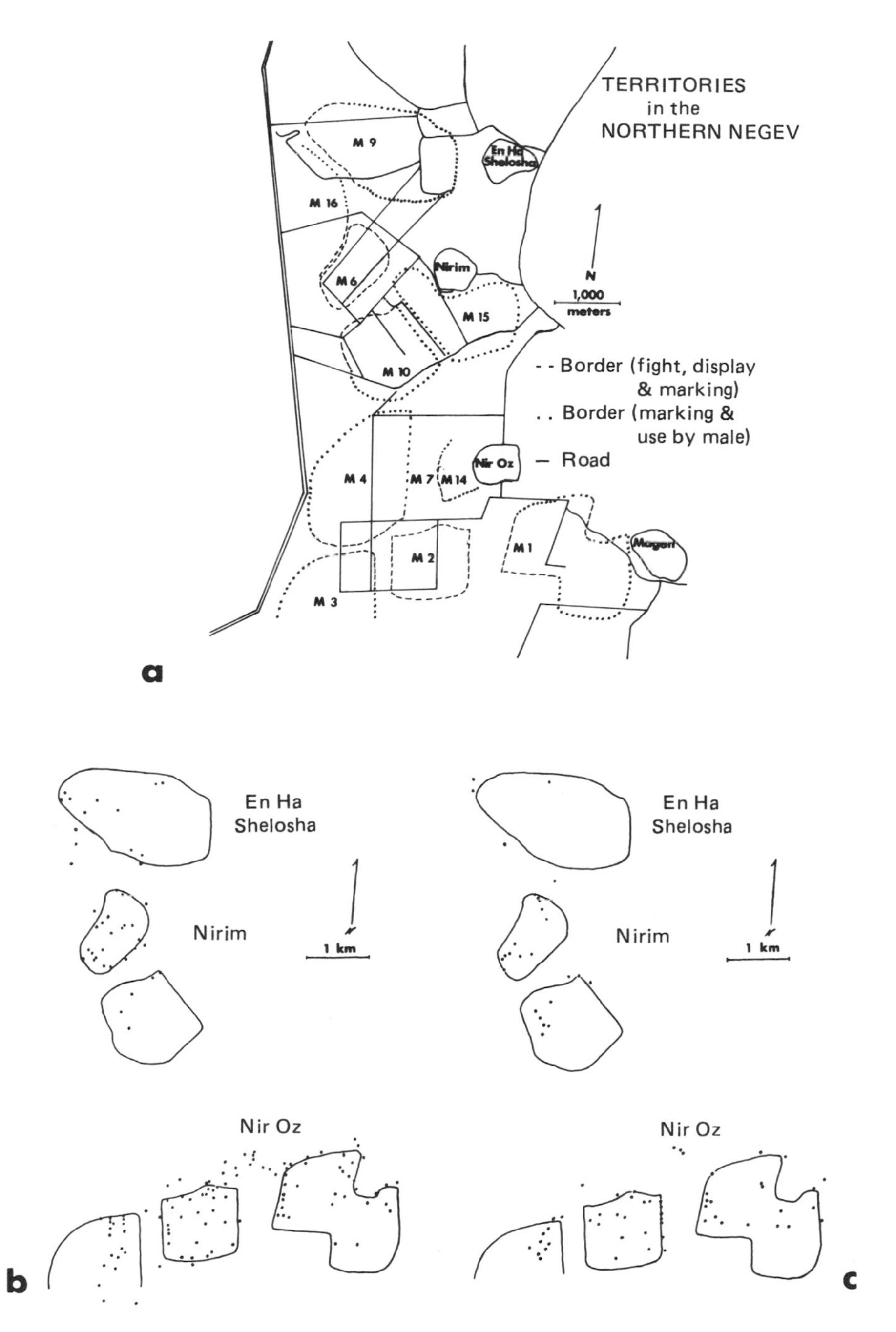

Figure 21: Map of mountain gazelle territories in northern Negev. (a) Borders of territories. (b) Urination-defecation sites in relation to territorial borders. (c) Object aggression sites in relation to territorial borders.

Territoriality in the mountain gazelle appears to be a more permanent situation than e.g., in Thomson's gazelle. The actual length of time a male is territorial is not known since the males observed either were already territorial when the study began, or were still holding their territories when the study ended. Thus, only fractions of the territorial periods could be recorded. These ranged from one to 20 months with an average of 13.5 months in 13 territorial males (Table 3). The two males with the longest territorial periods (each of them 20 months) did not show any signs of leaving their territories when the study ended. Determination of the extent of territoriality among the adult males was difficult due to the bachelors' large home ranges. The bachelor herds' normal wanderings took them away from the northern Negev study area for extended periods of time. When the different bachelor groups frequented the area at the same time, the proportion of bachelors in the adult population was high. Thus, the proportion of bachelors in the adult male population varied from 0% to 69%.

Indian Blackbuck

Sizes of the majority of blackbuck territories (Table 3) were calculated similarly to those of tommy males, the number of millimeter squares intersected on a map of known scale being counted. However, paced distances were used where necessitated by greater topographic relief, lack of adequate maps or very small territory size.

The sizes of 33 Texas territories ranged from 1.2 ha to 12.8 ha with an average of 4.1 ha (Mungall 1979). The minimum sizes of just more than 1 ha were the same in both medium-size and large pastures. In a pasture where large expanses of trees and brush tended to isolate the openings from each other, the largest territory (3.3 ha) only approximated an average-size territory in a pasture where large openings communicated with one another so that the bucks had the option of spreading out more while still having neighbors. On the other hand, of the two largest territory sizes observed for areas with more than one territorial male (11.3 ha and 12.8 ha), the one was in a large pasture and the other in a medium-sized pasture with approximately equal population densities.

The maximum limit for blackbuck territory size seems to approach 20 ha. A particularly aggressive territorial buck who managed to monopolize a large share of the one big opening in his medium-size pasture held 18.7 ha for more than a year before former bachelors began becoming territorial. The only adult buck in a large pasture with a few females and juveniles had an area of 19.6 ha in which he acted as if it were a territory with vague boundaries.

Two representative territories paced in Point Calimere Wild Life Sanctuary, where territories are scattered through a network of interconnecting openings, approximated the larger Texas territories for situations with more than one territorial male present. The two Indian neighbors held 12.4 ha and 15.1 ha. Near Mudmal (Andhra Pradesh) in India, Prasad (1981) measured 12 territories ranging from 3.3 ha to 16.6 ha with an average of 9.2 ha. These figures compare favorably with the average-to-large-size Texas territories. The situation at India's Velavadar Blackbuck National Park was quite different. Instead of a patchwork of openings and brush, flat open plains characterize the park. Here a tight cluster of about 25 territorial males (Figure 22) use a 19 ha corner of the park while large bachelor associations, female groups, and mixed herds roam the rest of the 3,467 ha and beyond into fields and scrubland. Two representative territories within the territorial mosaic measured 0.34 ha for a central location and 0.37 ha for a fringe location. Thus, the location of a territory in a central or a peripheral region of such a cluster does not seem to have striking effects on its size.

Figure 22: Exceptionally dense territorial mosaic in an "arena" of blackbuck males in Velavadar Blackbuck National Park, India. (Photo: E.C. Mungall.)

Obviously the Velavadar measurements are much smaller than the minimum sizes noted in Texas where tightly clustered territorial mosaics like that at Velavadar have never been reported. Also, in India, such small territories are by no means the rule. K.S. Dharmakumarsinhji (pers. comm.) has seen another case of such a cluster of small territories at the temple Sandhida Mahadeo. He is inclined to relate the formation of such dense territorial mosaics to the extra protection afforded by humans in these areas. One may also speculate that the extreme density of territories in a territorial mosaic like that at Velavadar may be a question of landmarks. Lacking striking ground features, the bucks rely more exclusively on their own "landmarks." Their conspicuous coat, displaying, and calls help them remain spaced from one another while their dung piles help them orient to fixed locations. A further condition favoring dense territorial mosaics among blackbuck may be huge herds like those of several hundred to more than 1,000 common at Velavadar. Isolated territorial males may be more prone to being swept away when large herds pass.

In any case, the system of tightly grouped territories works at Velavadar. Of 11% of the adult male population judged to be territorial, only 2% held places apart from the one territorial "arena." The percentage in the territorial mosaic could have been higher before visitors' lodge and park roads took half of the area traditionally used by the territorial bucks (Dharmakumarsinhji pers. comm.). Also, more bucks may seek territorial status later in the year during reproductive peaks. At Point Calimere where the territories are more spread out and where "big" groups are hardly a tenth as large as at Velavadar, 38% of the adult males appeared territorial. Here the season was more advanced and a reproductive peak was in progress during the study period.

In Texas where several populations have been observed regularly for one to two years, each large pasture always had at least some territorial bucks. However, summer tended to be the season when fewest bucks were territorial. Some of the territorial bucks observed took up their territories in the late summer or early fall; they held them throughout the winter and into the spring when they finally left to spend the remainder of the year in bachelor associations before resuming their territories. Nevertheless, prime space was seldom vacant at any time of the year. Bucks tended to return to the same territories after a period of absence as bachelors. A buck who failed to regain a territory formerly held might take up residence close by. These points are demonstrated by the diagrammatic maps for the ownership changes during January 1974 through January 1975 in one valley (Figures 23 and 24).

Some males centralize their activity in one area for long periods of time. Especially in smaller pastures with only one opening suitable for territories, one very dominant individual—as noted above—may take over practically the whole space and be the sole territory owner. This relationship can persist for well over one year. Even in large pastures with a number of openings, however, it is possible for an individual to keep a territory through all seasons. Another situation in which a buck may keep the same area for extended periods is that in which there are only a few females in a large pasture.

Furthermore, there are localized shifts within pastures. For example, some bucks will make a brief attempt at holding a territory again in another place not long after having left or lost a well-established holding. Bucks may also shift from one spot to another not far away or expand into adjacent areas. Normally, a buck who has given up his territory stays with a bachelor group for a while before becoming territorial again. However, two males changed location abruptly without being observed to revert to bachelor status in between. Prasad (1981) observed comparable events in India.

In six cases, both the beginning and the end of a territorial period were observed (Figure 25). In these instances, the territorial periods ranged from not quite 2 weeks to about 11 months with an average of approximately 4 months. In 14 further cases, either the beginnings or the endings of the territorial periods are not precisely known. These fractions included two periods of more than 17 months and seven periods ranging from between 6 and 11 months, suggesting that the average, from the relatively few fully observed periods, is probably somewhat too small. Near Mudmal in India, six out of nine territorial periods recorded lasted one to two months, but two lasted longer than eight months (Prasad 1981).

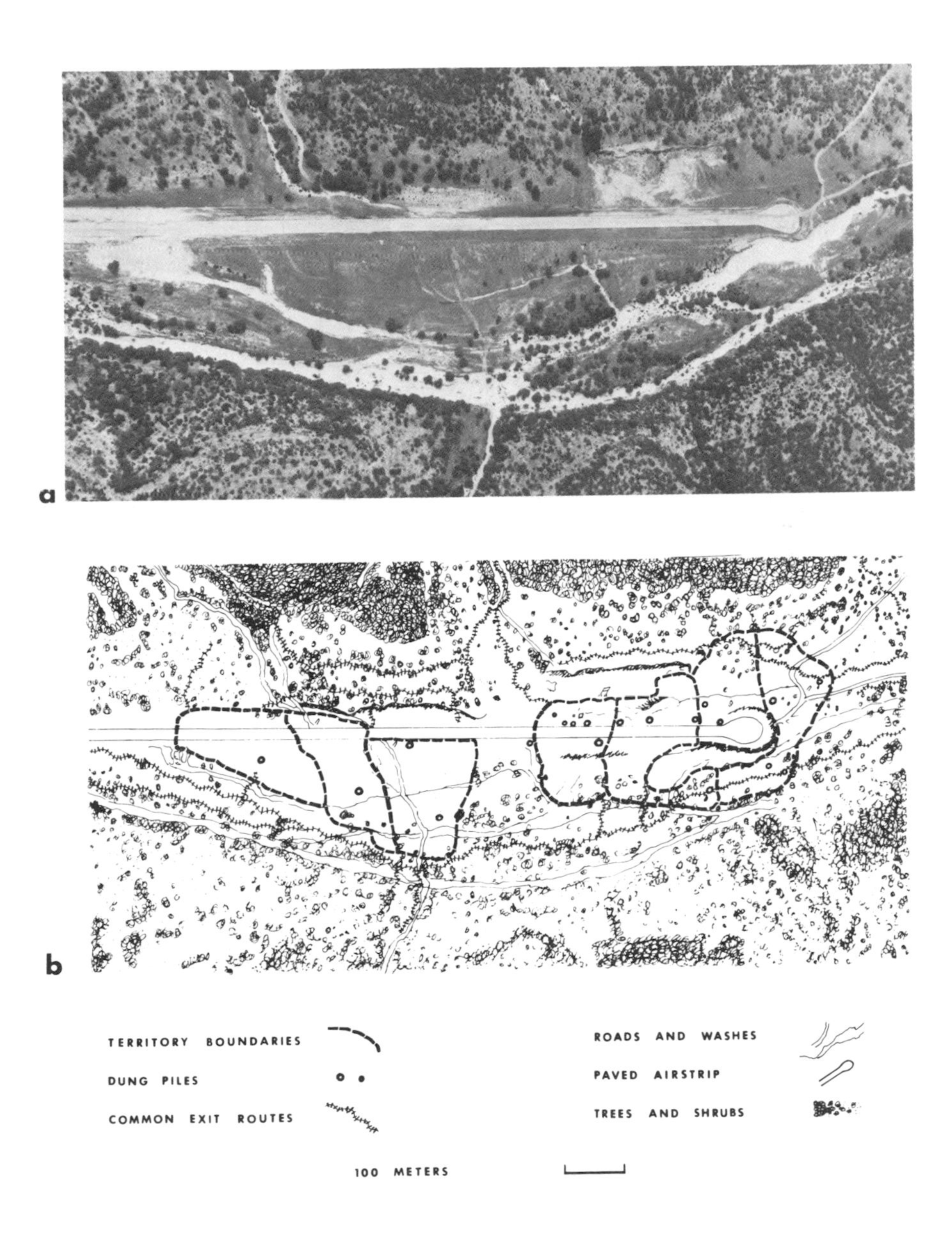

Figure 23: Map of blackbuck territories at a Texas ranch. (a) Aerial view of a valley at Greenwood Valley Ranch, Texas. (Photo: E.C. Mungall.) (b) Map of blackbuck territories in this valley.

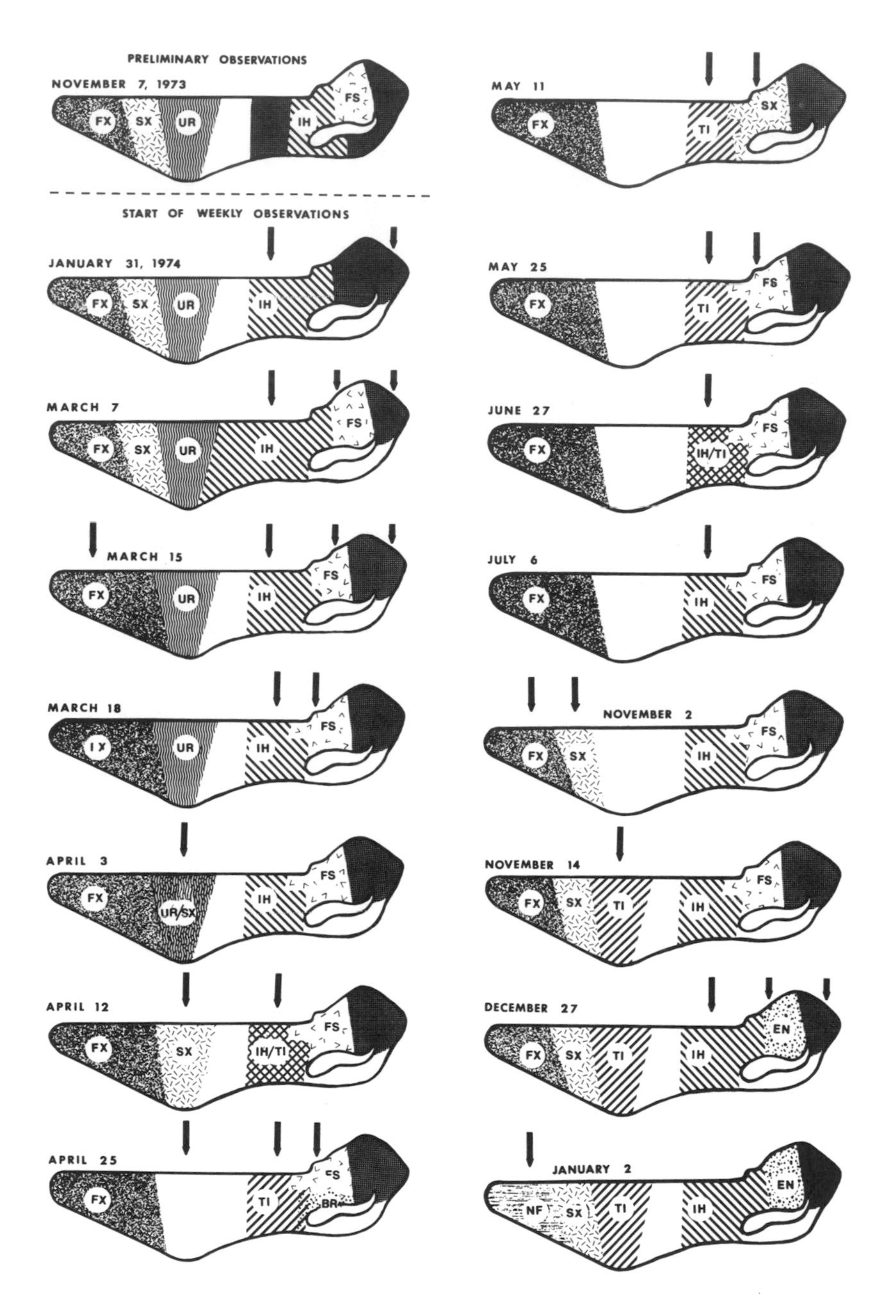

Figure 24: Diagrammatic maps giving territory changes along the valley shown in Figure 23 during one year. Blank areas were unoccupied. Two-letter designations identify known bucks who owned the territories. Where there is no designation, the owner, or succession of owners, did not include any of the known males.

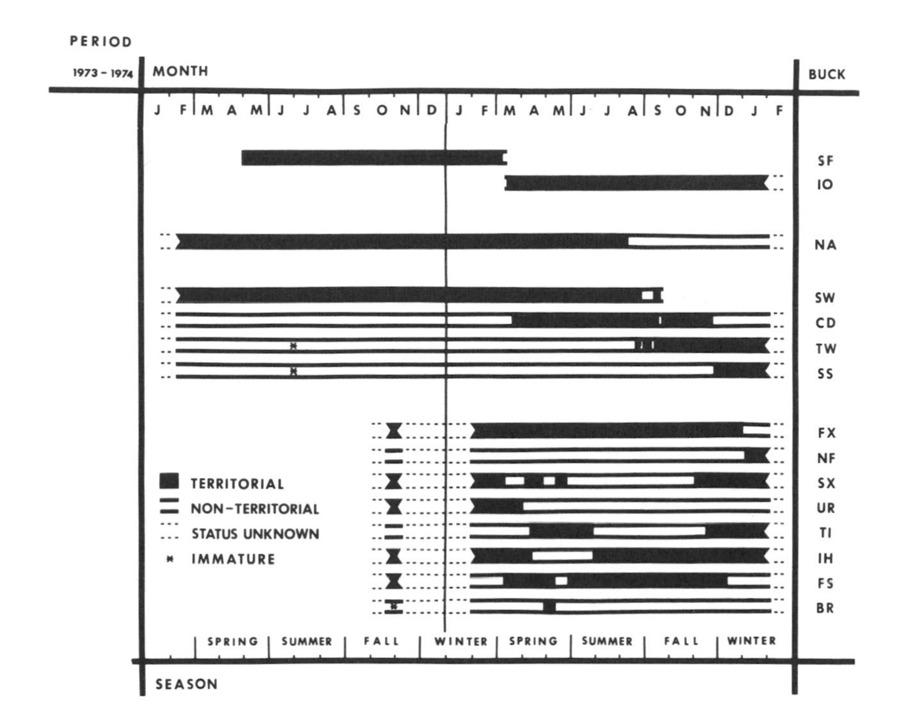

Figure 25: Some of the territorial periods of blackbuck males in Texas pastures.

Other Antilopinae Species

In gerenuk, Leuthold (1977, 1978a) did the most intensive studies on territorial behavior. In his observation area in the Tsavo National Park, Kenya, all the single adult males appeared to be territorial. In the two observation areas, these single adult males made up 10.3% and 18.1% of the population. Although Leuthold (1977) himself points out the necessity of distinguishing between territories and home ranges in ungulates, he apparently does not feel quite comfortable with this distinction in gerenuk since he alternately speaks of "home ranges" and of "territories" or even of "home range/territories" in these males. In 10 home range/territories investigated, he found sizes from 130 to 340 ha (Leuthold 1978a). For five individually known bucks, the territorial periods lasted 16, 17, 27, 30, and 41 months, i.e., all these territories were held for more than a year, with a maximum of almost three and a half years (Table 3). Small-scale shifts of territories toward, and away from, the Voi River were observed several times, apparently due to season.

Bigalke (1970) found that the numbers of solitary, presumably territorial springbok males increased during the rut. For example, in the Etosha National Park, solitary adult bucks made up 26.2% of the sighted groups during the rut in July, but only 5.9% outside the rutting season in April. In the Jack Scott Nature Reserve in Transvaal, Mason (1976) mapped three springbok territories with sizes of 27, 50, and 70 ha. He suggests lengths of territorial periods from four to 12 months, or possibly even longer (Table 3).

5

Environmental Requirements for Territoriality in Single Species

THOMSON'S GAZELLE

Environmental requirements are not the primary factors for territoriality even though they can influence location and structure of a territory as well as the length of the territorial periods. Perhaps Thomson's gazelle with their well-pronounced migrations in the Serengeti provide a particularly good example to make this point clear. First, territoriality is not limited to a definite season in this species, but occurs all the year long, although environmental conditions may drastically change with season. Secondly, some territories are always held even when the majority of the non-territorial conspecifics have left an area due to unfavorable conditions (see p. 184). Thirdly, Thomson's gazelle must traverse areas (e.g., relatively dense forests) in the course of their migrations which obviously are unsuitable for the establishment of territories in this species. Nevertheless, some males try to establish territories there although they cannot hold them for any length of time. In short, a tendency to become territorial is always present (at least in some individuals) regardless of environmental conditions.

Feeding conditions set only the lower limits. Of course, a buck cannot hold a territory where feeding conditions are entirely insufficient. Furthermore, poor feeding conditions may prevent females from entering the area. Although tommy males often establish and hold territories without females being present, it seems to be quite important that females visit them, at least now and then. When an area is not frequented by females for a long period, quite a number of the males abandon their territories. This seems to be a response to the absence of females, and, thus, only in some cases and even then only indirectly a response to feeding conditions. Vice versa, good feeding conditions may attract many females and make them stay in an area. Consequently, many males may establish territories there and keep them. In short, good feeding conditions in a given area can provide favorable conditions for territoriality, but they are not decisive for its occurrence.

Very similar is the situation with drinking. During the dry season, Thomson's gazelle generally need to drink, and they frequent open waters. However, territorial males apparently can tolerate long periods without drinking. As previously mentioned, one of the individually known males kept his territory for 11 months, i.e., also during the dry season. There was no open water in his territory, and he was seen only twice outside his territory during all this time, in both cases at locations without water. Even when this male was kept under uninterrupted observation for 48 hours during the peak of the drought, he did not leave his territory for one minute, and thus, he did not drink.

The most important environmental factors for successful territoriality in this species are open short-grass plains and dry ground. Thomson's gazelle do not establish territories on muddy ground or in high-grass areas. As mentioned above, tommy males may sometimes try to become territorial in woodland or brush areas. However, they regularly abandon such "experimental" territories after a very few days, if not after a few hours. At least one tree or bush inside the territory is comforting to tommy bucks when available. They stand in its shade during the hottest hours of the day. Probably, the availability of shade is also the reason why on a *mbuga* the open areas adjacent to the fringe of the woodland are preferred and are occupied first when Thompson's gazelle invade such an area. If no tree is in a territory, but is in the vicinity, the desire for shade may be one of the very few reasons why even a "good territorial" tommy buck occasionally leaves his territory. However, a shade tree is not absolutely necessary, and there are many tommy territories without any tree or shade.

Landmarks such as single trees, termite mounds, rocks, "isles" of high grass, dry creeks, trails, etc., are often used to establish the center and/or boundaries of a territory. The buck incorporates them, and the position of the boundary is often more precisely determined in the vicinity of such landmarks than in sections where no landmarks are available. However, the establishment of a territory does not depend on landmarks in this species, and by far not all available landmarks are incorporated. The male uses only those which happen to be in the immediate vicinity of the boundary as he "intended" it. Landmarks which do not coincide with the "intended" boundary line are ignored. Furthermore, tommy males are well able to establish territories in the absence of any striking landmarks. In such cases, the neighbors in a dense territorial mosaic as well as the dung piles and the secretion marks of the owner himself apparently provide a sufficient orientation to determine the position of the center and the territorial boundaries.

GRANT'S GAZELLE

In principle, Grant's gazelle also prefer short-grass plains for the establishment of territories, but they are not as confined to them as are Thomson's gazelle. If no short-grass areas are available or vacant, Grant's males can become territorial in areas either partially or completely covered with relatively high grass or scattered woodland. On the other hand, Grant's gazelle appear to be more dependent on landmarks, possibly because, in contrast to tommy males, they do not mark objects with their preorbital glands, and their territories do not form such a dense mosaic. Apparently, the landmarks do not necessarily have to be inside the territory. A striking landmark such as a *kopje* or a single big tree which can be seen at a distance, is sufficient for their orientation. However, one will hardly find a Grant's territory in an area where no such landmarks are visible.

MOUNTAIN GAZELLE

Territorial mountain gazelle males in the northern Negev did not regularly or even frequently leave their territories during their daily activities. Thus, the territory must contain enough food to fulfill the nutritional requirements of the male as well as those of the females that graze on his territory. Mountain gazelle rarely drank in the northern Negev, and so the bucks did not have to leave their territories for drinking purposes. The regular and rather heavy dew in this area probably provided adequate moisture. The gazelle on Ramat Yissakhar, with its hot and dry winds during the summer, were observed to drink from the stream in the valley. For this reason, the territorial males had to leave their territories on the slopes.

Mountain gazelle occupy a variety of habitats in Israel. They have adapted well to the habitats manipulated by modern agricultural methods and have successfully established territories in these areas. The territories under observation were either in areas with bare soil (plowed fields) and short herbaceous vegetation in gently rolling terrain or in areas with short herbaceous vegetation and steep slopes. Although gazelle entered the citrus orchards on the study areas, these orchards only formed a boundary of the territory and were excluded from the territory proper. Shade is not a requirement for mountain gazelle territories. However, when shade was available, the territorial males and the female herds occasionally rested in the shade.

In the northern Negev, mountain gazelle preferred habitat with little or no tall vegetation, and the growing agricultural crops presented a problem to them. As the crop height increased to more than 20 cm, the gazelle left the areas they had been using and entered areas with shorter vegetation (Table 4). When the wheat or other crop was harvested, they returned to their old areas of activity. Increased vegetation height also forced two territorial males to change the location of their activity centers. When the vegetation was cut, these males returned to their previous "core areas." In short, mountain gazelle clearly preferred to establish their territories in open country with low vegetation.

In the Yissakhar valley, the territories were located on the slopes of the valley. Due to the rough terrain, neighbors frequently could not see each other. In this situation, landmarks seemed to be important, and the boundaries of the territories followed the steeper parts of the slopes or the ravines that run down the slopes. Thus, such a territory usually was on relatively level ground (benches or topographic terraces) bordered by much steeper relief. The territorial boundaries in the northern Negev also followed the much gentler slopes in the few cases where slopes were available. The abundance of natural and man-made landmarks on both study areas precluded any possible natural experiment with the absence of landmarks.

Table 4: Northern Negev Vegetation Height During the Crop Growing Season in Relation to the Use of These Areas by Mountain Gazelle

Month	Vegetation Height and Gazelle Use: Bare Soil % Area	Bare Soil % Use	0–20 cm % Area	0–20 cm % Use	21–30 cm % Area	21–30 cm % Use	31–40 cm % Area	31–40 cm % Use	41–50 cm % Area	41–50 cm % Use	>50 cm % Area	>50 cm % Use	N Gazelle-Day* Observations
January	69	58	25	40	4	2	0	0	0	0	2	0	206
February	28	29	34	69	5	0	25	2	4	0	4	0	532
March	30	72	0	0	21	27	8	<1	1	0	40	<1	292
April	27	62	11	29	18	2	<1	0	<1	0	43	6	187
May	26	53	11	42	18	3	<1	1	1	0	44	1	506
N Gazelle-day observations	869		717		102		18		0		17		1,723

*One gazelle-day is one gazelle observed one or more times on specific census dates. (Percentages sum to 100 horizontally.)

INDIAN BLACKBUCK

Both in India and in Texas, blackbuck congregate on the open stretches of flat to gently rolling country for grazing. Where undisturbed, large herds form in early morning and again in late afternoon as small groups graze farther into the open. In these areas, the blackbuck's chief food, the grasses, grows in greatest abundance. This is where a buck has the best chance for coming in contact with the largest number of females in the shortest time. Since the blackbuck territories are reproductive territories, it is best for a male to occupy a flat, open, grassy area such as the females prefer to select.

The greatest hazard for the blackbuck during India's monsoon is flooding (Johnson 1975, Oza 1976, Dharmakumarsinhji 1978). Animals may be swept away or may die of exposure. In Texas, territorial bucks start returning to their territories as soon as flood waters recede. The changes they find in the arrangement of rocks, tree limbs and bushes leave the bucks unconcerned except at dung pile sites. All the dung piles are investigated and renewed. Territory borders are unaffected except for a loss in areas where rushing waters have widened gullies used as boundaries. Both defense and flight are laborious until the ground dries because mud and grass accumulate on the hooves of the blackbuck in balls weighing up to about 115 g.

The general preference of blackbuck for ground with low topographic profile is most pronounced with respect to territory sites. Males will use open slopes if the ground is fairly even, but horizontally flat land is the most preferred. This location also favors good visibility. The greater ease with which the bucks can use their principal courtship and herding display (p. 144) on flat terrain helps reinforce their preference for unobstructed, horizontal surfaces. Running fast with nose stuck up in the air becomes hazardous on broken ground. Similarly, the nose-up display impedes climbing. This disadvantage was well illustrated when a female which had been followed persistently by a male headed up a steep, rocky slope. The buck appeared in difficulty trying to climb the hill while maintaining his display posture. The female gained steadily. Finally the buck stopped, grazed, and returned to the valley.

Although blackbuck frequently inhabit lightly wooded country, bucks always select open ground for territories. Bucks may retreat into brush with the rest of their conspecifics for feeding to supplement what is available on their territories, to travel to water and when disturbed, but the territories are always clustered in the open. They tend to be so open, in fact, that some owners have no shade. If ambient temperatures rise uncomfortably high (above about 32°C) such an owner must leave his territory if he is going to spend his rest periods in the shade. While some leave, other may elect to stay and lie in the sun.

Blackbuck are one of the most sun-tolerant of the Indian ungulates. The Thar Desert has always been one of their strongholds. Any disadvantage caused by lack of shade in a territory seems far outweighed by the preference for expanses with good visibility. Only widely scattered trees, clumps of trees or brush are tolerated in a territory. For example, on one Texas ranch, a randomly selected northeast to southwest line transect crossing three adjacent territories as well as the space between the first of these and the next territory indicated that the last two territories were 100% open, that the first territory was 85% open, but that the unoccupied area was only 57% open.

On the other hand, the territories investigated in this study did not include any stretches of bare ground or unvegetated gravel even when available. Exceptions were dirt roads, ditches or strips of pavement which crossed some territories. A marked tendency to use these as boundaries is shown only when they are well above or well below the adjacent ground surface. For the most part, vegetative cover in the open areas consists of several short to mid-grasses with an admixture of forb species. If forbs growing thickly in a large part of a territory approach 1 m in height, that part is abandoned until these plants wither, break apart and thin out again. Clumps or narrow bands of tall annuals cause no shifts. The bucks merely add them to the objects they thrash with their horns and mark with their preorbital glands. Choosing open ground allows the bucks good visibility which they take advantage of to join females or to chase intruders as soon as either come close.

As mentioned above, boundaries do not necessarily coincide with paths, roads, ditches or other features which can be crossed easily. On the other hand, major discontinuities such as cliffs, thick stands of trees and deep or rocky water courses often become boundaries. Likewise, certain man-made structures—notably fences—may be taken as limits on one or more sides. Neither periodic flooding which shifts the arrangement of rocks and tree limbs nor the progressive growth and withering of patches of taller plants seems to change the border. Thus, as in Thomson's gazelle, blackbuck males may readily incorporate available landmarks when establishing territorial boundaries, but the environmental landmarks are not primary factors for territoriality.

OTHER ANTILOPINAE SPECIES

Springbok follow the usual gazelle pattern in that their territories are in open grassland. In areas with dense shrub savanna, Bigalke (1972) found territorial males on small pans which provide miniature islands of open country in the otherwise dense brush. These pans are partly bare and support a sparse vegetation of grasses and karroid shrubs. Each one is occupied by a solitary adult male.

That territories are in open areas with very low vegetation and even partially bare ground, may also be presumed for other gazelle species such as dorcas gazelle or Loder's gazelle, mainly because this type of landscape makes up the greatest part of the habitat of these species. However, even when they have a choice, a preference for open areas is likely. For example, a captive male goitered gazelle kept in an enclosure of somewhat more than 1 ha with half of it covered by conifer forest of the middle-European type and the other half an open meadow, established a territory on the meadow up to the fringe of the forest. Except for days with extremely bad weather, he was hardly ever seen in the wooded half of the enclosure.

Gerenuk males establish territories in relatively dense vegetation for an Antilopinae species. Leuthold (1978a) describes his study area in Tsavo National Park as essentially flat country consisting mainly of light to moderately wooded grassland and open bush or woodland.

6

Structure of Territory

RESTING SITES AND OTHER SPECIFIC ACTIVITY SITES

Some animals give a certain structure to their territories or home ranges simply by restricting definite maintenance activities to specific localities. They may have definite feeding areas, watering places, bedding sites, trails, urination and defecation spots, etc., within the territory (Hediger 1949). This structuring of a territory by localizing maintenance activities seems to be rather limited in Antilopinae.

Usually, Antilopinae territories do not include any open water. Thus, the possibility of special drinking sites within a territory does not exist in most cases. Except for temporal restriction of (all) activities to core areas—as discussed e.g., for mountain gazelle (p. 72)—grazing and browsing take place wherever the corresponding vegetation is available. Therefore, feeding is not restricted to definite sites. Grooming and scratching are frequent after a rest and may be seen more frequently in certain sections of a territory than in others since bedding sites can be localized in a territory (see below). However, this clearly is a secondary effect. In principle, scratching and other types of grooming are independent of locations.

It is possible and often even likely that the movements of the owner follow a daily routine within a territory. However, these movements do not usually result in a highly visible trail system. Of course, there can be trails in a territory, and they also can be used by the owner. However, these trails come from afar, cross the territory, and continue beyond it. Thus, they obviously are established by other animals, but not by the territorial male.

Urination and defecation are largely localized in territorial Antilopinae males, but they are so closely linked to the establishment of a marking system that they need to be discussed in this context. The other maintenance activity which can be bound to definite sites within a territory, at least in some of the species under discussion, is resting in both forms, standing and lying.

No definite resting sites were recognized in territorial mountain gazelle males (Grau 1974). In the publications on territoriality in springbok (Bigalke 1970, 1972, Mason 1976), this point is not mentioned. Also, in gerenuk, Leuthold (1978a) mentions only the exposed standing of territorial males on termite mounds (Figure 26) which may be related to thermoregulation, antipredator behavior, and/or static-optic advertising (Hediger 1949). This type of standing, of course, is restricted to definite locations. Grant's and Thomson's gazelles were also observed very occasionally standing on elevated ground, such as termite mounds or small hills. In free-ranging Grant's gazelle as well as in dorcas and goitered gazelle in captivity, preferred bedding sites were noticed within the territories, but were not investigated in detail. Thus, some detailed information is presently available only for Thomson's gazelle and Indian blackbuck.

Figure 26: Territorial gerenuk buck standing exposed on termite mound. (Photo: W. Leuthold—Tsavo National Park, Kenya.)

As previously discussed, territorial bucks in these species like to spend the hottest hours of the day standing in the shade of a tree or bush. However, due to their strong preference to establish territories in open areas, if trees are available at all, usually only one to three are within a territory. Thus, standing in shade is restricted to these very few places in the territory. Furthermore, territorial tommy males often spend remarkably long periods standing at a definite spot right on the boundary, predominantly on boundary sections where agonistic encounters with neighbors are frequent. This standing is probably a combination of static-optic advertising and watching the surroundings, apparently not so much for predators but more for rivals and female herds.

One to three definite bedding sites are in the center of a tommy territory. Some more are close to the boundary, again especially in those boundary sections where interactions with conspecifics are frequent. Altogether, a tommy buck may have up to about 10 bedding sites in his territory. A preference for them is clear. Nevertheless, this preference does not absolutely exclude resting at other places.

Territorial blackbuck males frequently lie right on their dung piles (Figure 18c), and, thus, on very definite spots within their territories (Mungall 1978a). This habit appears to be rather unique among Antilopinae. At least, it has not been observed in any other Antilopinae species up to now. It also is not frequently found in other bovids, e.g., it has been reported from blesbok (*Damaliscus dorcas phillipsi*—Lynch 1974).

CENTER AND BOUNDARY

Geometrically, of course, each territory has a center. The question is whether the central region has a special meaning to the owner. For several reasons, this question is not easy to answer in Antilopinae, and the answer becomes more difficult with larger territories. For example, in birds, the nest may characterize the center of a territory, or in rodents and certain carnivores, it may be a den. In bovids such as wildebeest *(Connochaetes taurinus)*, which roll on the ground as a comfort behavior, a territorial male may have an obvious wallowing place in the central area. Antilopinae territories lack such striking features of the center.

In Antilopinae, the majority of interactions (display encounters, fights, herding, etc.) take place in the boundary regions of a territory. Consequently, one may expect the central area to be the place for more "peaceful" activities and for relaxing, above all for resting. At least in the small territories of Thomson's gazelle, this expectation is true. As mentioned above, one to three resting sites are found in the approximate center of a tommy territory, and one of these usually is the most preferred resting place within the whole territory.

Early ethological literature (e.g., Tinbergen 1953) sometimes gave the impression that the so-called "defense" by the owner against intruders became more frequent and more vehement as the intruder invades deeper into a territory, and thus, the center would be the most strenuously "defended" area of the whole territory. At least for Antilopinae, this picture is wrong. Here, full and vehement "defense" occurs as soon as an intruder crosses the boundary. Thus, most of the agonistic encounters take place in the boundary region. Exceptions may occur, e.g, when a territorial buck temporarily tolerates subordinate males within his territory (see p. 184), or when he, for some reason, did not become aware of an intruder until the latter arrived in the central region, etc. However, these are well defined, special cases which are relatively rare.

In addition to "defense," there are several more behavioral symptoms listed previously (p. 6) which indicate the existence and the position of the boundary in an Antilopinae territory. Perhaps with the exception of gerenuk (Leuthold

1978a), it is therefore not too difficult for a human observer to become aware of the boundary and its location. Finally, the system of dung piles and/or secretion marks reflects the position of the boundary as well as that of the center, at least in some species, as we will discuss soon.

Depending on several factors, the boundary can be a sharp line, or it can be a broader zone, a belt, a "no-man's land." Generally, the larger the territory, the larger is the boundary zone. For example, in mountain gazelle with their very large territories, a breadth of the boundary belt of 50 to 100 m is not unusual, but in blackbuck with its much smaller territories, the boundaries are strips only some 15 to 20 m in width. When a territorial buck has incorporated striking landmarks, such as single trees, termite mounds, etc., into the boundary line, or when he has established dung piles there, the boundary can be determined to the very meter in the vicinity of these landmarks or dung piles. However, it may be a broader belt in those sections where landmarks or dung piles are absent, or, at least, are not incorporated into the boundary line. Furthermore, the presence of neighboring territories considerably helps define the boundary. The more frequent agonistic interactions with territorial neighbors are, the more precise is the boundary line. Likewise, the position of the boundary is often very sharply delineated in those sections where non-territorial males frequently invade a territory or where females frequently cross when entering and/or leaving the territory.

Such favoring factors can add together, e.g., when a striking landmark is incorporated into the boundary line, and the male has established a dung pile there, and another territory is adjacent, and females frequently pass there. In such a case, e.g., the border zone of a Grant's territory can dwindle down to only a very few meters in width even though the boundary belt commonly is about 30 to 50 m broad in the large territories of this species. When such favoring factors are present in territories of small size, as in Thomson's gazelle, cases of very precisely determined boundary lines become quite frequent. On the other hand, when the opposite factors add together, e.g., when no neighboring territory is present in a boundary section, landmarks are not incorporated into it, and no dung piles are established there, or only a very few at large intervals, etc., then even a tommy territory may have a relatively broad boundary zone.

MARKING SYSTEM

For the structure of a territory, particularly for the subjective establishment of a center and a boundary, the owner's marking activities are of special importance. All the marking behaviors—such as urination-defecation sequence, scraping the ground with the forehooves, deposition of gland secretion, and object aggression—are also performed by non-territorial adult males. However, they occur more frequently in territorial individuals, and the activities of the latter are related to the structure of the territory. Thus, it is not the marking behavior per se which is characteristic of an individual's territorial status, but the enrichment of a limited area with the marks of this individual and the system resulting from their spatial arrangement and distribution (Walther 1964a).

Aggression toward inanimate objects, i.e., vehement rubbing of horns and forehead against trees or bushes, thrashing grass or goring the ground with the horns, is perhaps the most questionable means of marking. Of course, one could argue that such object aggression may leave visible spoor in the environment indicative of the presence of the performer. On woody vegetation, the marks of object aggression often are quite conspicuous but in herbaceous vegetation, they are hard to detect for a human observer—which, of course, does not exclude the possibility that the animals may be able to detect them. Some authors (e.g., Brooks 1961, Grau 1974) have also considered the possibility that gland secretion may be transmitted to vegetation during object aggression. This appears to be somewhat unlikely in most Antilopinae since, to date, no skin glands on the foreheads of these animals have been described, and secretion from the preorbital glands deposited in this way has never actually been detected.

In mountain gazelle (Figure 21c), five to 19 object aggression sites, usually in herbaceous vegetation, were found in five territories investigated. Most of these sites were in the broad boundary zone and in sections where encounters with neighbors or bachelors were frequent. In other species, the numbers and positions of the object aggression sites were not investigated in detail due to the mentioned difficulties in detecting them in grassland vegetation.

Unfortunately, not much is known about the role of the secretion from a number of skin glands in territorial marking. For example, some of the species under discussion have inguinal glands and/or carpal glands. Secretion from these glands is probably deposited on the ground when the animals lie down for resting, and may give a specific odor to the bedding sites. Likewise, it is possible that secretion from the interdigital glands may be deposited when the animals paw the ground with their forelegs, in addition to visible spoor resulting from this behavior. However, no definite proof has been presented for all these possibilities up to now, and they remain more or less speculative. Also, the role of urine in the marking of Antilopinae has not been investigated in detail. Thus, what remains to be discussed is marking the territory by dung piles and by secretion from the preorbital glands.

Dung Piles

Dung piles have been found in the territories of all Antilopinae species investigated up to now (Figure 27). However, the numbers and positions of the dung piles within a territory have been described in relatively few cases. For example, in gerenuk and springbok, it is only known that there are dung piles in the territories; small ones in gerenuk, big ones in springbok. Captive dorcas gazelle established only three dung piles in an enclosure of about 1 ha; one of them in the center. In contrast, a captive male goitered gazelle, raised by bottle and strongly imprinted to humans, established a whole chain of dung piles with short intervals along that section of the fence which separated him from human visitors. Thus, the territorial boundary was particularly well marked in this case.

In mountain gazelle, dung piles are found mainly in the vicinity of territorial borders (Figure 21b). Most of the dung piles are located on the slopes instead of on top of the low ridges in the northern Negev. They are also found beside the roads and at field edges. The dung piles along field edges occur primarily when the tall vegetation of one field limits the activity of the buck in that field and thus forms a temporary edge to this territory.

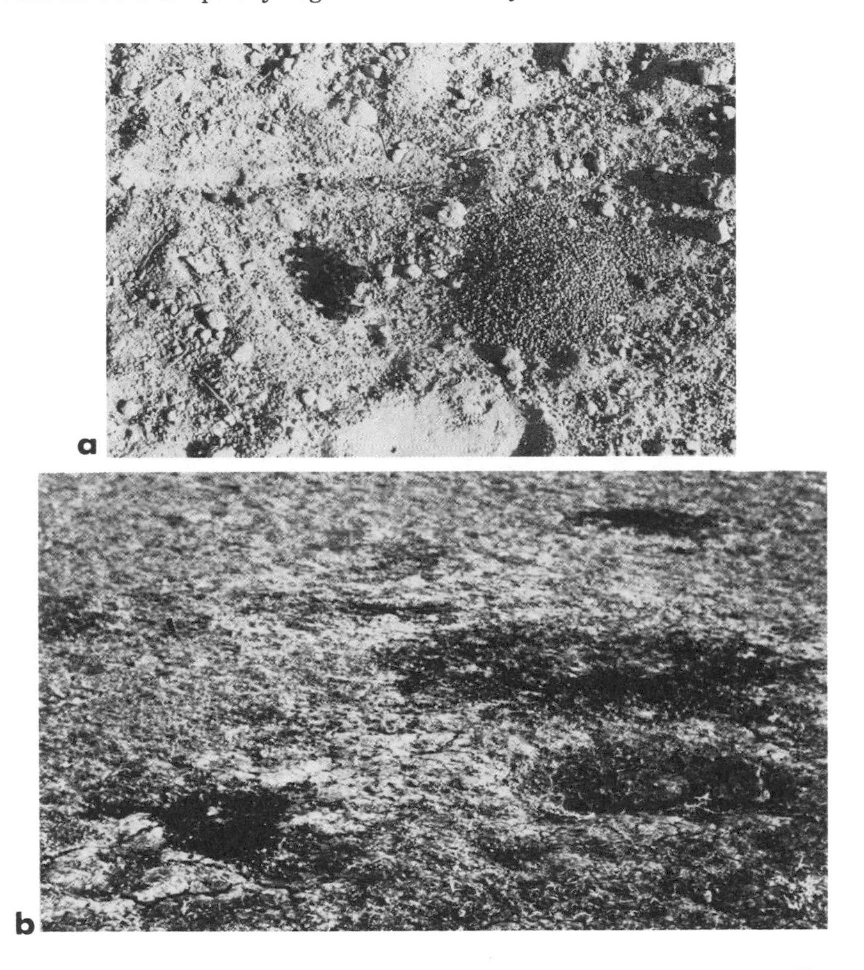

Figure 27: Dung piles of territorial males. (a) Dung pile of mountain gazelle showing scrape marks and fresh dung. (Photo: G.A. Grau—northern Negev.) (b) Clustering of dung piles in dense territorial mosaic of blackbuck. (Photo: E.C. Mungall—Velavadar Blackbuck National Park, India.)

On the whole, the marking systems of territorial mountain gazelle in the northern Negev do not appear to be highly structured. The large territories and wide border zones may partially account for the loose dispersion of dung piles. There also appears to be a tendency for more frequent marking near borders where territorial neighbors are known to occur. Some dung piles are used by both neighbors. Territorial males frequently make special trips to go to a dung pile. After urinating and defacting there, they return to their original activity in whatever part of the territory they had come from. In six mountain gazelle territories investigated, the number of dung piles ranged from five to 34, with an average of 20. These dung piles were rather small as compared to the fewer but larger dung piles on Ramat Yissakhar where they were not destroyed by cultivation. Scraping the ground with the foreleg frequently preceded the urination-defecation sequence. This scraping tended to scatter and enlarge the dung piles on Ramat Yissakhar and left highly visible marks in the bare soil in the northern Negev.

In Indian blackbuck, the territories have well defined boundaries and dung piles (Figure 27b). Dung piles may also occur outside of territories in this species, but the active dung piles a buck has within his territory are of special significance to him. He maintains them as stamps of his ownership and special behavioral sequences are associated with them. Other males do not recognize his boundaries unless he is there to enforce them, and in his absence strangers treat his dung piles as they would any dung piles outside territorial boundaries. There is at least one dung pile within every territory. One or more are often situated well inside with another or others near the edge. Those at the periphery are close to where the buck occasionally leaves or enters his territory either when beginning or ending interactions with other blackbuck or when going to or from points outside the territory for food and water. When re-entering the territory or when concluding an encounter with another buck, the owner goes to a dung pile, lowers his head, sometimes circles with head still low, scrapes with a foreleg, may perform *Flehmen,* urinates and defecates (Figure 28). In its briefest form, all but the urination and defecation in extreme postures are omitted. Then the owner either grazes away from the dung pile or lies down on it.

Bucks in general check the olfactory condition of any dung pile they come across. Strange bucks and bachelors wandering by a dung pile pause and lower the nose to it, often exhibiting *Flehmen* (see p. 141) before or as they move away. Although bachelors also use dung piles, they do not establish them; creation of dung piles is the prerogative of territorial bucks and territorial hopefuls.

A territorial blackbuck can locate the exact sites of his dung piles even after a flood or a bulldozer has removed the fecal pellets and disturbed the surface soil. As when a dung pile has been experimentally manipulated or when a stranger has used a dung pile, the owner investigates and then proceeds to renew the dung pile with a temporary elevation in its frequency of use.

Blackbuck dung piles tend to be round or oval. When distinctly oval, the long axis reflects the most common direction which the buck takes when entering onto or leaving the dung pile. Dung piles on trails are often oval for the same reason. Maximum lengths were 4.7 m (N=138) on the Edwards Plateau of Texas, 2.0 m (N=49) at Velavadar in India, and 1.4 m (N=51) at Point Calimere in India. Lengths of about one meter were standard at all sites with the Velavadar dung piles, which were on the hardest ground, averaging the largest. At all three sites, many of the dung piles showed signs of scraping to the extent that they were dished below the surrounding ground surface. Sandy Point Calimere had the highest proportion of dished dung piles (86%) as well as the deepest dishes (about 8 cm as opposed to about 5 cm in the other localities). Velavadar with its hard soil had the lowest proportion of dished dung piles (33%). Even on oval dung piles, dishes tend to be round. The dishes, which are the site of most

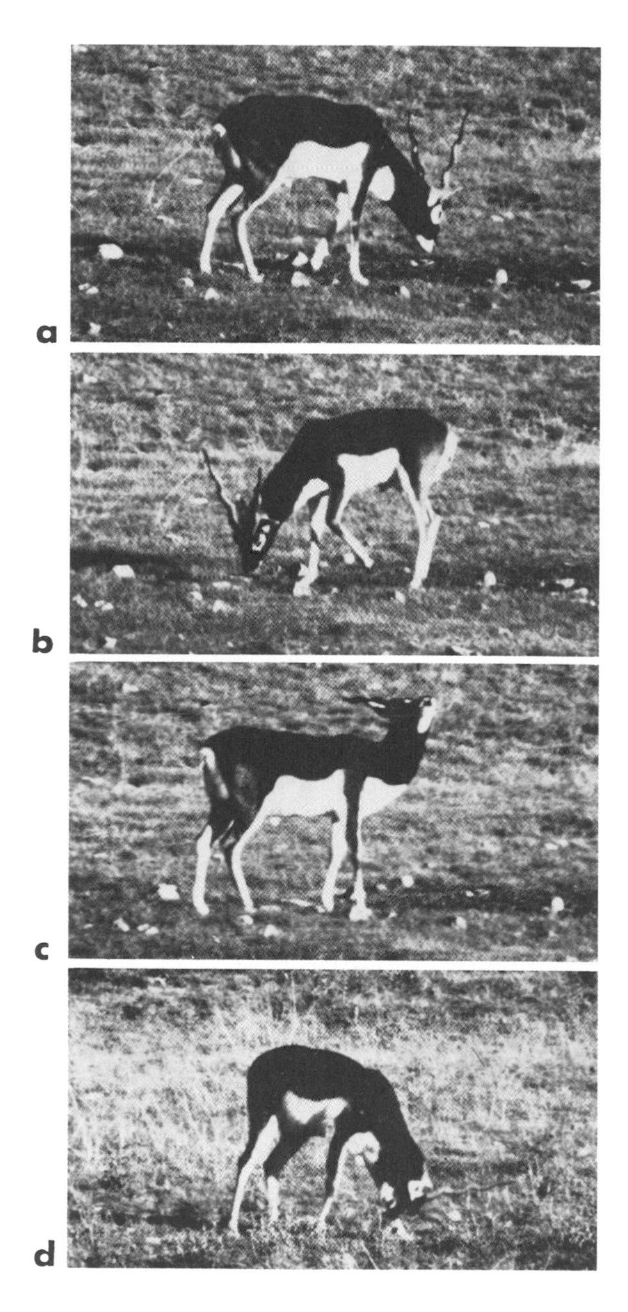

Figure 28: Behaviors of a territorial blackbuck prior to urination and defecation at a dung pile. (a) Sniffing at the dung pile (typical beginning). (b) Scraping and circling with lowered head (frequent interlude). (c) *Flehmen* (common). (d) Thrashing (occasional). (Photos: E.C. Mungall—Greenwood Valley Ranch, Texas.)

active deposition, usually are only a quarter to a third as long as the total dung pile. The flat area of the dung pile surrounding the dish represents mainly pellets raked out of the dish by pawing and further strewn when a buck lies on them. Therefore, although the buck does frequently lie on his dung pile rather than merely beside it, he is not necessarily lying on freshly deposited pellet clumps because he does not typically lie on the dished part.

In seven blackbuck territories investigated on Texas ranches, the number of dung piles per territory ranged from two to nine, with an average of five. Always one dung pile was in the approximate center of the territory. Of the remaining 29 dung piles, 24 were in the boundary region. At Point Calimere where average territory size is more like maximum size in Texas, maximum dung pile density as indicated by a mapped sample of 1.2 ha was 14 per ha. This number was bracketed by sample values for the fringe of the territorial mosaic at Velavadar and for the central portions: three per ha on fringe and 40 per ha near center. In Velavadar's territorial mosaic, territories are so small that nearly three can be accommodated on 1 ha of ground. Two basic tendencies probably contribute to the sharp increase in dung pile density through the central region of the dense territorial mosaic. First, a territorial buck frequently maintains a dung pile right across the border from an adjacent neighbor's dung pile. Although by no means obligatory, this tendency for paired dung piles is still evident in Texas pastures. Second and more subtle, a new owner taking over a territory previously occupied by another buck sometimes creates a new dung pile of his own next to that of the last owner. He may do this only at a major dung pile within the boundaries he has inherited or he may do this at one or more lesser dung piles as well. If a previous owner later retakes the territory, he renews his former dung piles and lets the new ones paired with them fall into disuse. The conspicuous clusters of dung piles noted at Velavadar are interpreted to be the extreme of these tendencies as encouraged by the extreme in small territory size in an area highly preferred for territories (Figure 27b).

In Grant's gazelle, the large territory size poses difficulties in counting the number of dung piles an owner has. Also, their frequent location in areas with higher vegetation than in mountain gazelle, Thomson's gazelle or blackbuck complicates the task. Furthermore, the large Grant's territories frequently include or overlap territories of other species such as Thomson's gazelle, topi and kongoni, and all these other animals establish dung piles, too. Although it is usually possible to distinguish the droppings of Grant's gazelle from those of the other species, occasionally there are problems.

All the dung piles of a territorial Grant's male were detected, with a reasonable degree of assurance, in only two cases. In one case, there were six dung piles, with one of them in the center and four of them on the boundary, within a territory of about 60 ha, and, in the other case, four dung piles, with one of them in the center and two on the boundary, in a territory of about 45 ha. In all the other observations, there was always at least one, but usually several, dung piles in a territory; however, no evidence was found that their numbers ever considerably exceed the maximum (about 10) of dung piles in the much smaller blackbuck territories on Texas ranches.

In Thomson's gazelle where 17 territories (Table 3) were investigated in this regard, one to three dung piles are in the approximate center, and a "chain" of

10 to 20 goes along the boundary of a well established territory (Figure 29). Of course, it takes a buck at least one to two weeks to establish such a dung pile system. In the beginning of a territorial period, only a few dung piles, or even only one, are found in his territory. If a male abandons the territory before he has reached the peak of territoriality, he may never establish a complete dung pile system.

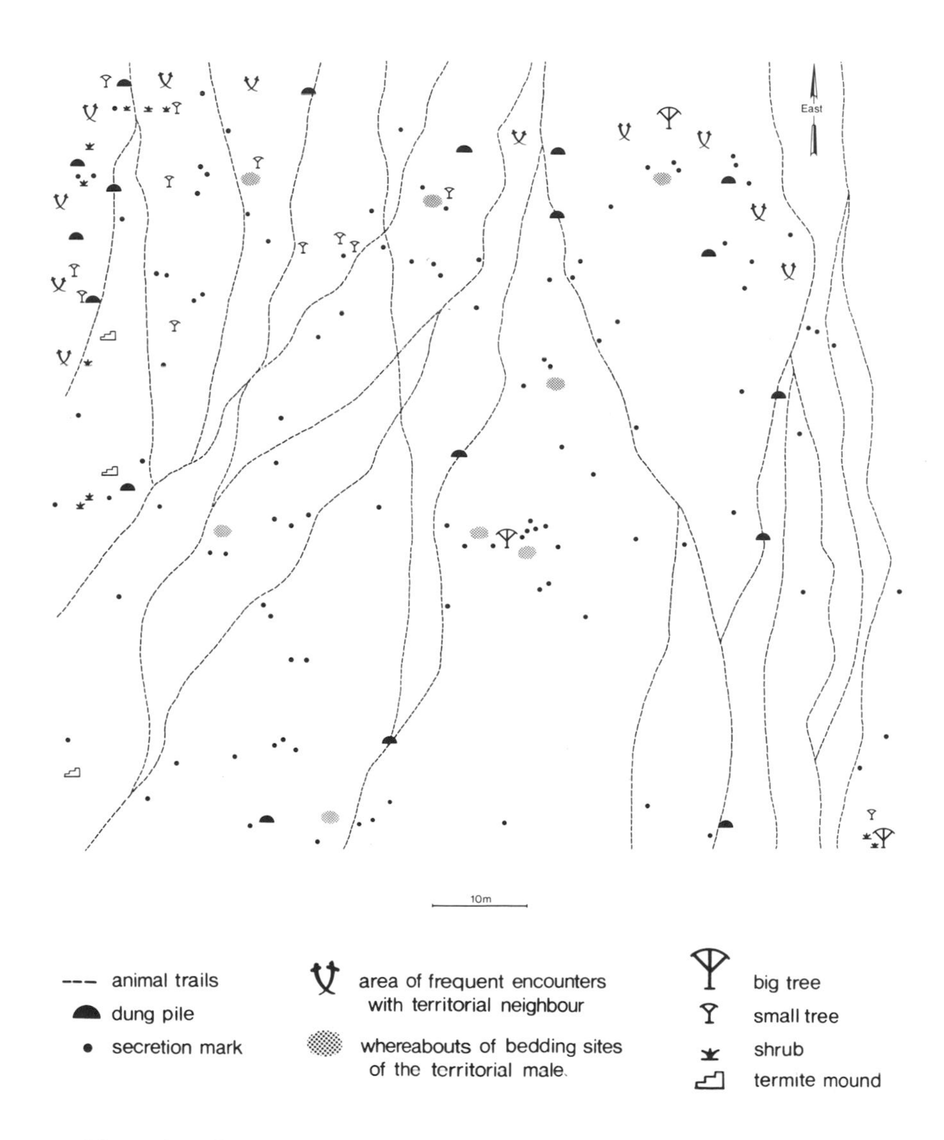

Figure 29: Structure and marking system in a territory of Thomson's gazelle. (For details of mapping procedure see Walther 1978c.)

In a tommy territory, dung piles are more frequent in boundary sections where agonistic encounters are frequent (Figure 29). Boundary dung piles usually are established by only one male. However, occasionally two territorial neighbors may use the same pile, or a buck may establish a dung pile very close to a boundary dung pile of his neighbor. Some dung piles always are found near preferred bedding sites. Some may be on trails, vehicle wheel tracks or roads, as far as they cross the territory.

The establishment of dung piles on trails and roads seems to result from a tendency observable in many ungulates (Walther 1978c, 1979) to establish dung piles on bare ground. The dung piles close to preferred bedding sites are certainly linked to urination and defecation after a long rest which is very common in many animals. Since at least one preferred bedding site is in the approximate center of a tommy territory, also one or more dung piles are located in a central position.

Furthermore, animals frequently defecate in "exciting" situations. When a buck is right on the boundary and close to leaving his familiar territory, he may easily become "excited." In a few instances, territorial tommy males even showed the same sequence of conflict behaviors when temporarily leaving their territories as described for subordinate males when threatened by superiors (Walther 1969). This "excitement" may partially account for the dung piles along the boundary. Agonistic encounters with neighbors certainly create "excitement," and urination and defecation are frequently observed in this context. Thus, dung piles are particularly numerous in those boundary regions where hostile interactions take place.

After dung piles have been established in a territory due to defecation close to bedding sites and/or at occasion of agonistic encounters, later, as a routine is established, the territorial male tends to restrict urination and defecation to these places, i.e., he also defecates there when he has not encountered an opponent and when he has not just been resting. One may say that the dung piles per se are now attractive to the male, and he goes there primarily to urinate and to defecate. This tendency to place droppings at a spot where there already are some, is widely distributed in animals (Altmann 1969); however, it apparently is developed to different degrees in different species and sometimes even in animals of the same species according to their sex, age, and social status. In Thomson's gazelle, young, females, and non-territorial adult males usually urinate and defecate at random. The mechanisms described here for Thomson's gazelle, probably are more or less equally valid for the establishment and the locations of dung piles in other gazelle species. In Indian blackbuck, females and young urinate and defecate at random, but adolescent males begin to use dung piles, older bachelors ordinarily do and territorial males do almost without exception.

Secretion Marks from Preorbital Glands

Although all the gazelles and their relatives possess preorbital glands, only some of them mark environmental objects. In contrast to certain other bovid species, such as blesbok and klipspringer, the females do not mark at all in Antilopinae. Immature males also do not actually mark, but they may already perform the marking movements at a time when apparently the preorbital glands are not yet functioning (Walther 1973b). Even adult bucks do not mark objects in all the Antilopinae species. For example, Grant's gazelle, Sömmering's gazelle, dama gazelle, springbok, dorcas gazelle, and mountain gazelle do not mark, but Indian blackbuck, gerenuk, red-fronted gazelle, Thomson's gazelle, goitered gazelle, and probably also dibatag do mark. The situation within the species *Gazella gazella* is interesting. The mountain gazelle in Israel *(Gazella gazella gazella)* were not observed marking objects during two years of intensive study. However, Indian gazelle *(Gazella gazella bennetti)* in the Zoological Garden in Karachi, Pakistan, were seen marking the fence of their enclosure. Thus, it appears that there even are differences among the subspecies within the same species in this case.

In most of the object marking species named above, no special studies have been conducted concerning the marking system or the locations of marks. Thomson's gazelle is the single species in which some detailed information is presently available, not least because the secretion marks are easy to see in this species. Even here, the marking system in only one territory has been investigated, and the smallest (0.8 ha), complete, and functional territory available was selected because one almost literally must look at each blade of grass in such an investigation. (For technical details of mapping the marking system see Walther 1978c).

One-hundred-ten secretion marks were found in this territory. (In another, incompletely investigated territory, 65 secretion marks were found—Estes 1967, Walther 1964a.) Twenty-four of the 110 marks were in those boundary sections where agonistic encounters were frequent, 11 were in the immediate vicinity of bedding sites in the center of the territory, two to five in the vicinity of each of the other bedding sites, and the others distributed on the boundary "belt." The boundary belt of secretion marks was broad in all sections with adjacent territories but formed only a thin line along the western boundary of this territory where no neighbor was present ("open boundary"). Except for the immediate vicinity of bedding sites, the central area of the territory was weakly marked, almost unmarked.

In a tommy territory (Figure 29), the secretion marks follow about the same distribution pattern as the dung piles; there are marks in the immediate vicinity of bedding sites (and thus, often also close to dung piles) and along the boundary. Particularly when threatening an opponent before a fight and later during the grazing ritual (see p. 118) after the fight, territorial tommy bucks frequently mark with their preorbital glands. However, the concentration of secretion marks in boundary sections of frequent agonistic interactions is not as striking as that of dung piles. The secretion marks in the boundary region form a much broader "belt" as compared to the "line" of dung piles. This distribution appears to be linked to the habit of territorial tommy males to mark a grass stem, then walk a few steps and mark another grass stem; to walk, to mark, etc. Particularly in the early morning, a territorial buck often takes a "marking walk" during which he may mark up to 15 times with intervals of a few steps, apparently renewing part of his marking system.

Functions of Olfactorial Marks Within a Territory

Schenkel (1966), who has paid particular attention to territoriality and marking behavior in mammals, has put emphasis on olfactorial marks, such as urine, dung piles, and secretion from skin glands. He even suggested restricting the term "marking behavior" exclusively to olfactorial marking. His argument was that, at least in mammals, only the olfactorial marks are independent of the producer and, thus, are present and effective even in his absence. Other authors have not gone that far. However, the opinion is widespread in ethological literature that such olfactorial marks advertise to conspecifics the occupation of a given area by a territorial individual, that these marks have a repellent effect upon other males, and thus, that they constitute a means of "territorial defense."

There is very little evidence for this view in either Antilopinae or other territorial ungulates. Occasionally, a bachelor entering a territory may sniff at a dung pile or secretion mark, and then turn around and leave. However, usually the non-territorial males do not hesitate to enter well-marked territories, and the intruders do not show signs of being intimated by the olfactorial marks. Thus, dung piles and secretion marks do not prevent invasion by non-territorial males and are not a means of territorial defense in Antilopinae. Females also do not pay noticeable attention to the olfactorial marks, and thus, the assumption that these marks may attract females is highly unlikely.

A certain importance of dung piles and secretion marks to the owners of adjacent territories is not as unlikely as a repellent effect upon bachelors or an attraction of females. However, also in these cases, specific reactions are rare and anything but striking. Even when an experimenter switched droppings from the dung pile of one territorial blackbuck to one of his neighbor's, the latter only showed moderate excitement for three minutes after coming to the manipulated dung pile for the first time. He sniffed at it, scraped and then also thrashed in two bouts, urinated and defecated twice, and exhibited *Flehmen* once. Head-low circling came intermittently between these actions. Thus, he only somewhat deviated from the usual in intensity and frequency of activities commonly exhibited when a male visits a dung pile. Another territorial blackbuck completely ignored cotton-tipped sticks coated with preorbital secretion from bachelors. Likewise, when in Thomson's gazelle, grass stems with thick secretion marks resulting from multiple marking, were cut and planted in a densely marked section of the neighboring territory, the buck detected some of them, sniffed at them, marked over them with his own secretion and continued his routine walk. His behavior was not different from that in another experiment in which thin wooden sticks without any secretion were planted in the same area. It appeared that the male's behavior was more a reaction to something new within a very familiar area than a specific reaction to a territorial competitor. On the other hand, it is likely that a territorial buck is aware of the position of his neighbor's dung piles and "belt" of secretion marks, and thus, these olfactorial marks may somewhat contribute to a better recognition of the boundary by territorial neighbors.

On the whole, olfactorial marks may not have a great function in territorial advertising to others, but they may be more important to the owner himself. They certainly emphasize the subjective structure of a territory, and in this way, they may facilitate the owner's orientation within it. Furthermore, the enrichment of a limited area with his own marks and their odor, may familiarize the owner with this area, make him "feel at home," and, thus, may enhance his self-security in that area.

7

Male Behavior at Peak of Territoriality

In the following discussion, behavior patterns performed by a male at the peak of his territorial period, i.e., after the territory has been well established but before signs of a decline in territoriality become evident, are described. We must emphasize at the start of this discussion that almost none of these behavior patterns are exclusively restricted to territorial individuals. This statement includes only a minor exaggeration in that a very few of the behaviors under discussion were not, or only infrequently, observed in non-territorial adult males. For example, copulation has, to date, only been recorded in one case involving a clearly non-territorial Antilopinae male (blackbuck), and this case was absolutely exceptional in several regards. Adult bachelors sometimes pay attention to females, and one could imagine that the sexual approach of a non-territorial buck may occasionally end with copulation. On the other hand, copulation is only a special case of mounting, and mounting definitely also occurs in non-territorial males. Futhermore, a few vocalizations given while herding females or chasing bachelors presently have so far been described exclusively from territorial males. However, one could imagine that they exceptionally may be uttered by a non-territorial male. The grazing ritual of Thomson's gazelle (see p. 118) has definitely been seen in non-territorial adult bucks; however, it is extremely rare among them, whereas it is the usual ending of a fight between territorial neighbors. Thus, roughly speaking, it can be used as an indication of territoriality in this species.

All the other behaviors occur in territorial as well as in non-territorial adult males. The urination and defecation postures are often more pronounced in territorial individuals, and usually all the behaviors under discussion are more frequent in the territorial bucks than in the non-territorial males (see p. 165). Some behaviors such as "pushing" a female also occur in bachelors, but in territorial bucks, they are related to the structure of the territory, e.g., when a female is herded away from the boundary toward the center of the territory. On the whole, however, it certainly is an important and interesting fact that territoriality has no (or only a very few and somewhat dubious) special movements or vocalizations which exclusively belong to it.

MARKING BEHAVIOR

Urination-Defecation Sequence and Scraping the Ground

Before urinating and defecating, an adult Antilopinae male usually lowers his head and (presumably) sniffs the ground or the dung pile. Only blackbuck sometimes circle in this situation, and occasionally exhibit *Flehmen* after sniffing the dung. Then still with lowered head, the Antilopinae bucks paw the ground or the dung with a foreleg, but this pawing is not obligatory. After scraping, the male steps forward stretching his body and lowering his belly toward the ground, the hindlegs sloping backward. He urinates in this posture (Figure 30a). Immediately after urination, he abruptly brings his hindlegs forward so that their hooves are placed lateral to the stationary forelegs. He assumes a crouched posture with his anus only a few centimeters above the ground and defecates (Figure 30b). In this way, urine and feces are deposited on the same spot. The hair of the white rump patch is somewhat fluffed, and the tail is lifted to a horizontal position, or, in blackbuck, it is curled over the back. Only the dibatag male sometimes beats his long tail between his hindlegs during urination, apparently wetting his tail with urine in this way. After defecation, the male resumes a normal posture, wags his tail in some species, and walks off or starts grazing. Blackbuck males also may bed down at the dung pile.

Particularly in young males, it may sometimes happen that the male urinates and then walks a few steps before he defecates. According to Grau (1974), linking the urination and the defecation into a sequence generally is infrequent in non-territorial mountain gazelle males. In other gazelle species, the sequence seems to be obligatory in adult males. Even when such a male occasionally defecates without urination, he almost always assumes the urination posture first; one may say as a vacuum activity. Urination and defecation may take place at any time; however, they are particularly frequent after a longer rest and in connection with agonistic encounters, as discussed above. A captive dibatag male's urination-defecation sequence was sometimes also released by the female's urinating or defecating (Walther 1963a). The buck approached the urinating female, sniffed at her urine and performed *Flehmen*. Then he pushed her rump with his horns and forehead to move her away. He scraped, urinated and defecated placing his urine and droppings on top of hers (Figure 31). Meanwhile, she usually returned, sniffed at his urine and performed *Flehmen* on her part. No comparable "ritual" is known from any other gazelle species. Also, it is unknown whether this behavior is generally typical of dibatag, or whether it was due to the special conditions of captivity (only one pair was kept in a relatively small pen).

The urination-defecation sequence has been found in all Antilopinae males investigated to date. The urination and defecation postures of males are so striking that almost all observers are inclined to regard them as visual displays. However, it should be emphasized that this assumption goes back to the

Figure 30: Urination-defecation sequence in a territorial mountain gazelle male. (a) Urination. (b) Defecation. (Photos: G.A. Grau—northern Negev.)

Figure 31: Urination-defecation "ritual" among male and female dibatag. (a) Female defecating. The male gently pushes her away. (b) The male urinates and . . . (c) . . . defecates on top of the female's droppings. (Photos: F.R. Walther —Naples Zoo, Italy.)

author's impressions and that it is not substantiated by observations of specific reactions by conspecifics. Exceptions are the cases of occasional contagion (social facilitation) as may happen with almost any behavior, i.e., when a male urinates and defecates and another buck is close by, the latter may also urinate and defecate. The most frequent case of this kind is simultaneous or consecutive urination and defecation of two territorial neighbors involved in a boundary encounter. On the other hand, the striking male urination and defecation postures can obviously emphasize the sex of the performer and may sometimes be suppressed when such an advertisement appears to be inopportune. For example, a young blackbuck male old enough to use the stereotyped sequence regularly when alone or with females frequently abandons the postures in the presence of an older male. Otherwise, the older male will go toward him displaying aggressively and force him to abandon the stereotyped postures and move ahead.

As mentioned above, scraping the ground or dung pile with a foreleg may or may not precede urination and defecation. The buck performs a few strokes with one foreleg and then, possibly, a few strokes with the other foreleg, but he does not strike alternately with both forelegs. Pawing the ground may sometimes be continued during urination in several species; in dibatag, it seems to be somewhat more frequent than in the other species. Frequently, but not always, the neck and the head are lowered during pawing.

The functions of scraping the ground before urination and defecation are not quite clear. On bare sandy ground, scraping leaves visible spoor. Thus, one could assume that a visual mark is added to the urine and the feces. However, this interpretation certainly can only be applied to a fraction of the cases. Another possibility would be that the male may add secretion from his interdigital glands to the urine and feces. The interdigital glands are opened at maximum when the two hooves of a foreleg are spread, as they are during scraping. On the other hand, the interdigital glands are located above the hooves, and, thus, only an abundant, "overflowing" secretion would reach the ground. But the secretion of these glands is not abundant in Antilopinae. In blackbuck and mountain gazelle, the scraping also aids olfactory checks of the droppings by disturbing the surface of the dung pile. In blackbuck, the incidence of pawing increases after any disturbance of the dung pile, and is also high during establishment or take-over by a new territory owner. However, in other species, pawing also occurs when no previous dung is present. In short, each of these interpretations of the function of foreleg scraping may contain some truth, but none of them offers a generally applicable explanation.

Scraping the ground also occurs before bedding down for a rest, in searching for food, and in connection with aggressive interactions (Figure 32). Before bedding down and in searching for food, it is much rarer than before urination and defecation in Antilopinae. Scraping the ground in agonistic encounters usually precedes urination and defecation in these species. However, there are rare cases in which pawing is not followed by urination and defecation during aggressive interactions. For example, in Thomson's gazelle, these rare cases still give the impression that the buck had urination and defecation "in mind" but that he did not get around to doing so because of the pressure of the situation. However, in Grant's gazelle, it happens that a male may scrape the ground during fighting, when his horns are interlocked with those of his rival. Again, such cases as these are rare, but they have been observed several times beyond any doubt. Of course, it is unlikely that urination and defecation were intended to follow in this situation. Thus, one may say that there are indications of a separation of scraping the ground in agonistic encounters from scraping the ground before urination and defecation.

Figure 32: Marking-related behaviors in a territorial dorcas gazelle. (a) Scraping the ground. (b) Object aggression. (Photos: F.R. Walther—Chaibar, Israel.)

Marking with Preorbital Glands

The following description of marking with the preorbital glands in Thomson's gazelle may be representative (Figure 33a) for all Antilopinae species in which the males mark objects. The buck frequently sniffs the grass stem before marking it. Occasionally he may also lick it or nibble at it. Then, he twists his head so that one cheek points more or less to the ground, and he opens both preorbital glands. He carefully lowers the preorbital region toward the grass stem

Figure 33: Preorbital gland marking in Thomson's gazelle. (a) Territorial male marking a grass stem. (b) Secretion mark on a grass stem. (Photos: G.A. Grau—Serengeti National Park, Tanzania.)

bringing the stem's tip into the widely opened gland. With quivering movements of the gland's margin, he deposits some (in the case of Thomson's gazelle, dark-brown) secretion which soon hardens in the open air (Figure 33b). After this procedure, he sometimes marks the same grass stem a second time with the same or the other preorbital gland. When a male stays in a territory for a long time, a specific grass stem may be marked quite frequently resulting in a pea-sized secretion pearl on its tip, sparkling in the sun. In blackbuck, whose secretion is not highly visible, some small hairs may be found on frequently marked objects.

The males may casually mark during feeding, or they may specifically approach a grass stem in order to mark it. The "marking walk" of territorial tommy bucks after day-break was previously mentioned. In territorial gerenuk males, Leuthold (1978a) found a significant increase in marking activity in late afternoon. Preorbital gland marking before or after urinating and defecating is quite common. Also, the marking activity frequently increases in connection with agonistic encounters, i.e., the males may mark before they begin to threaten the opponent, after a threat when the recipient is withdrawing, in a pause during a fight, and after a fight. In the last case, marking is particularly frequent during the grazing ritual of territorial tommy bucks. Finally, a male frequently marks before or after herding females and also before or after leaving or reentering his territory.

The marked objects are grass stems and blades, all kinds of forbs, branches, twigs of brushes and trees and, in captivity, also posts, fence wire, etc. Preferably, a buck marks objects of about his body height, but he also may mark objects that are as high as he can reach while still standing on all four feet. This height is the upper limit since Antilopinae males do not rise on their hindlegs for marking. When no objects of preferred height or above are readily available, as is frequent in open short-grass plains, males mark lower grasses and other vegetation down to a very few centimeters above the ground. Since the males prefer to mark the tips of objects, thin, sharp ends or edges strongly facilitate marking; however, it is not an absolute requirement. For example, it is not the case when a gerenuk buck marks a female by rubbing his preorbital region on her chest near the beginning of her neck, on her back at the croup, or on her rump (Figure 34).

More or less in connection with courtship, this marking of the female by the male has been observed in gerenuk in captivity (Backhaus 1958, Walther 1958) and in the wild (Leuthold 1978a), and in dibatag in captivity (Walther 1963a). The significance of this behavior is unknown. In other gazelle species (Dittrich 1965), males have been seen marking the tips of the horns of lying females in captivity. However, the latter has never been confirmed by observations in the wild. It probably is due to lack of adequate inanimate objects under captive conditions, i.e., the male uses the tips of the sharp, thin horns which are at about his body height in a lying female as substitutes for grass stems and twigs. Indeed, the need for marking can hypertrophy under captive conditions. For example, when rubber hose lengths were fastened at appropriate heights in a blackbuck enclosure with several males, they promptly concentrated on marking them (Mungall 1979). More dominant bucks even crowded against inferior males who were marking and hit them over the back with their horns to make room for themselves so that they could add their own secretion.

Figure 34: Gerenuk male marking a female with his preorbital gland. (Photo: W. Leuthold—Tsavo National Park, Kenya.)

Perhaps we should mention that the preorbital glands of the males also are wide open in both agonistic and sexual encounters. Of course, this behavior has nothing to do with object marking, but it may indicate an intensive emission of odor from the preorbital glands in these situations. When one takes into account how insignificant the effects of secretion marks are upon social partners, it is unlikely that the emission of secretion odor from the opened gland should have strong effects upon rivals or females. Probably, it works primarily as some kind of self-stimulation or reassurance of the producer which certainly is not out of place in such encounters.

Acoustical and Static-Optical Marking

There are no specific acoustical displays of territoriality in Antilopinae; however, there are vocalizations which sometimes may function this way. They occur when a male intensively herds females or when he vehemently chases inferior male opponents such as bachelors trespassing in his territory. Since herding females as well as chasing other males are frequent activities of territorial Antilopinae males, the vocalizations uttered in these situations may at the same time advertise the position and the status of a male, provided that they are loud enough. Most of the vocalizations uttered by Antilopinae males during courtship and during agonistic encounters are rather soft. The latter are even so soft that one usually can hear them only while playing the role of an opponent in encounters with completely tame animals in captivity. However, there are a few vocalizations which are loud enough to be heard at a distance.

In Grant's gazelle, a loud roaring from the open mouth was recorded only a very few times, always when females persistently tried to leave a territory, and finally were successful ("call of frustrated herding"). Much more frequent is a vocalization in Thomson's gazelle which was regularly heard when a territorial buck chased bachelors but sometimes also when he intensively herded females. It is a vibration sound uttered through the nose (as are most of the vocalizations in Antilopinae) and strophically repeated. As with many vibration sounds, it varies somewhat with distance and most commonly sounds like "pshorre-pshorre-pshorre" (o as in "off," e as in "left"). Similar but even louder is a vocalization "rόho-rόho-rόho" ("rocho" as first characterized by Walther 1959) in blackbuck. At each of the one to ten repetitions in the series, the sender expells a cloud of breath through open mouth and nods the head sharply. It is issued by an animal taking a dominant stance and almost always while in nose-up display (see p. 115). Consequently, it is typically a male call given while driving a female or when displaying to another buck. It often comes at the end of a chase as when the dominant buck slows and lets the fleeing subordinate run on, when a buck who has been absent reaches his own territory after having to "run the gauntlet" of surrounding terrain held by neighbors, or when a territorial buck on his dung pile senses a herd entering the vicinity. In this way, the vocalization seems to help advertise both the buck's presence and his mood (Cary 1976b). Although this vocalization occurs at all times of day, its frequency increases near sunset and decreases again after dawn. Partly this is linked to the increased activity during cooler hours, but poor visibility also seems to be a factor. During a 24-hour watch in northwestern India, no "rόho" calls were detected between sunrise and sunset, but 84 sequences were noted between dark and sunrise. Most came during moonlight activity. In mountain gazelle, too, "driving calls" are heard when a territorial buck chases females or other males.

Most striking are the vocalizations of territorial springbok when chasing bachelors or herding females (Bigalke 1972, Walther 1981). Although springbok have been said to be almost mute (Cronwright-Schreiner 1925), the territorial males are the noisiest of all Antilopinae species, and a good territorial mosaic of springbok males can be heard at a distance. There is a whole scale ranging from a repeated soft "brrl," through a considerably louder repeated "ru" (u as in "murder"), to a very loud "ru-oo" (oo as in "boot"). While approaching a trespassing bachelor or female, the territorial buck frequently intensifies his vocalizations throughout the whole scale.

Static-optical marking (Hediger 1949) does not require special displays in Antilopinae; it only requires that the territorial male be easily visible. In open short-grass areas where an animal the size of a gazelle can be seen from a great distance, the constant presence of a buck in a definite, limited area is sufficient for territorial advertising (Figure 18). Of course, this presence becomes more emphasized the more the male visually exposes himself with respect to the environment or conspecifics. Environmentally, he can expose himself by standing on a hill, a termite mound, etc. (Figure 26). Socially, he exposes himself when he does not intermingle with the females in his territory but keeps somewhat away from them (Figure 17a) so that he can easily be seen as a separate figure apart from the bunch of females. Emphasis of a territorial male's presence may also be achieved by placing himself on a spot of special structural importance in his territory and staying there for some time. In Antilopinae, the most striking example is provided by a blackbuck with his dark coat lying at a dung pile in his territory in open country.

Object Aggression

Pushing, beating, goring, etc., with the horns against inanimate objects has been observed in all the Antilopinae species under discussion (Figures 28d, 32b) and it has been described, in part, by different terms such as redirected aggression, object aggression, thrashing, and horning. The objects attacked can be bushes, small trees, branches, grass, the ground, and, in captivity, also fences, feeders, etc. Some authors are inclined to distinguish several categories of object aggression according to these different objects; particularly, they sometimes distinguish between horning vegetation and goring the ground. However, it seems more likely that a male just uses whatever object is most readily available to him under local circumstances and, thus, differences in objects do not mean much. For example, Leuthold (1978a) observed shrub-horning in gerenuk in the wild, but he did not see horning the ground. However, a captive male of this species who had neither shrubs nor vegetation in his pen, definitely did rub, butt and gore the ground with his horns (Walther 1961).

It is not easy to attribute a definite function to object aggression. Sometimes, one might think of play with inanimate objects or of a simple outlet of aggressive surplus energy. This interpretation is particularly likely in the cases when a male starts thrashing during grazing or browsing, i.e., when neither a rival is present nor any other reason for this aggression is detectable. Sometimes aggressive tendencies apparently are released by the presence of a potential rival, but the aggressor is more or less prevented from attacking him and fights an inanimate object as a substitute. Object aggression of this type, of course, is a clear case of redirected response (Moynihan 1955). In territorial Antilopinae males, such cases are particularly frequent when bachelors are in the surroundings but well outside the territorial boundaries. The aggression of the territory owner is released by the presence of these males but he would have to leave his territory to chase them away, and leaving conflicts with his tendency to stay on his ground. From such cases, it is only a small step to the use of object aggression as a means of threat, and sometimes object aggression clearly serves this function. However, among all the many threat displays of Antilopinae (see p. 113), object aggression seems the most inefficient because quite frequently the addressee ignores this threat or simply watches the performance without showing any reaction (Walther 1978a). As previously mentioned, object aggression may also leave spoor in the vegetation or on the ground, and, thus, could also be related to marking behavior. This relationship goes even further in that object aggression sometimes occurs before or after or intermittently with preorbital gland marking of the same object. Blackbuck rarely thrash vegetation without pausing to mark with the preorbital glands, and even mountain gazelle which do not mark with gland secretion frequently sniff the site before and after object aggression. In certain species, the movements of object aggression apparently have become ritualized so that they may well fall into Hediger's (1949) category of dynamic-visual marking. In some Antilopinae species, alternate pushing to the right and to the left has become an almost rhythmic performance which can be continued for several minutes or up to a quarter of an hour. This rhythmic "weaving" is most pronounced and frequent in Grant's gazelle (Walther 1965), but it has also been observed in Thomson's gazelle, dorcas gazelle, and mountain gazelle.

AGGRESSIVE BEHAVIOR

Fighting Techniques, Types of Fights, and Chasing

The terms "fight" and "fighting" refer here to any form of physical, agonistic contact. Severe fights and sparring matches are not differentiated because playful sparring is somewhat unlikely in territorial individuals, the principal techniques are the same in both serious and playful encounters, and the distinction between severe and playful fights is often more or less a matter of the observer's impression. Detailed descriptions of the fighting techniques and the forms of fighting in Antilopinae and other bovids have been given in several books and papers (summarizing reviews, e.g., in Walther 1968a, 1978a, 1979). Consequently, a relatively brief outline may be sufficient here.

Fighting Antilopinae males have never been observed to use forelegs, hindlegs, teeth, neck, etc. They fight exclusively with their horns. Attacks directed at the body are exceptional in most of the species under natural conditions. They may occur somewhat more frequently within the spatial limitations in captivity, and they seem to be more frequent in springbok than in other species (Walther 1981). Normally, the combatants fight horn to horn (Figure 35).

Figure 35: Horn-to-horn fight in adult mountain gazelle bucks. (Photo: G.A. Grau—northern Negev.)

This may be partially due to skillful maneuvers of the defender. However, gazelles and their relatives also usually *direct* their attacks toward the opponent's horns. Since the horns are the least vulnerable part of the other's body, one may speak of a ritualization of the fighting behavior toward conventional interactions which largely exclude bloodshed and fatalities. Most frequently, the fighters are oriented frontally toward each other, but orientation at an angle up to almost parallel position can also occur in horn-to-horn fighting. Antilopinae neither rise on their hindlegs (as e.g., goats frequently do) in fighting, nor do they drop down on their "knees" (carpal joints—as e.g., wildebeest, topi, kongoni, etc., routinely do). They fight while standing on all four legs with the forelegs widely spread in any fight of intensity, and the hindlegs in more or less normal position or relatively close together. Flat forward and backward leaps with all four legs are more common in some of the species (e.g., Thomson's gazelle) than in others (e.g., Grant's gazelle). In fights of great intensity, movements of the torso are also shown, mainly humping the back, stretching the torso, deep ducking of the chest region, and a vehement throwing of the whole body to the side (while horns are firmly interlocked). However, the leg and torso movements only support the actions of the head and neck which are the most important fighting movements for gazelles as well as horned ungulates in general. The following head and neck movements are recognizable in Antilopinae fights:

(a) *Nod-butt* (Figure 36a)—the animal pulls its chin somewhat toward its throat so that the horns point forward-upward. The opponent is hit with the upper third of the frontal side of the horns.

(b) *Forward-downward-blow* (Figure 36c)—is an intensification of the nod-butt. The movement of the head is the same, but the neck participates considerably more in this action than in the mere nod-butting. From a position of body level or higher, the head is brought forward and downward close to the ground by a vehement movement of the neck. The tips of the horns point upward and somewhat forward. When both opponents perform this simultaneously in frontal position, the frontal long sides of their horns clash together or their horns cross each other. A posture of readiness (Figure 36f), probably derived from the forward-downward-blow, is frequently assumed by Antilopinae males immediately before the beginning of a fight. The head is carried at body level, and the forelegs are widely spread. A variety of offensive or defensive fighting actions can follow this posture.

(c) *Push-butt*—In this fighting movement the posture of head and neck remains unchanged from the posture of readiness. The fighter only jumps or rushes forward. The horns of the opponents hit each other along the frontal surfaces or cross each other.

(d) *Forward-push and forward-swing*—The combatant pulls his chin toward his throat so that the tips of the horns point horizontally forward. The head is so deeply lowered that the forehead and the frontal sides of the horns almost or even literally touch the ground. From this posture, the animal pushes forward by stepping, jumping or rushing toward the opponent. The forward-push usually follows after another fighting movement, frequently after a forward-downward-blow. When both fighters simultaneously use the forward-push, their horns cross and interlock at the bases so that the opponents' foreheads almost or do in fact touch each other. When the head in the posture of forward-pushing has been placed so far backwards that it is between the fighter's forelegs, he may move the head forward and slightly upward in a long, flat arc. This forward-swing (Figure 36d) constitutes an intensification of the forward-push with the neck being strongly involved in the movement. It can also be performed when the horns are interlocked.

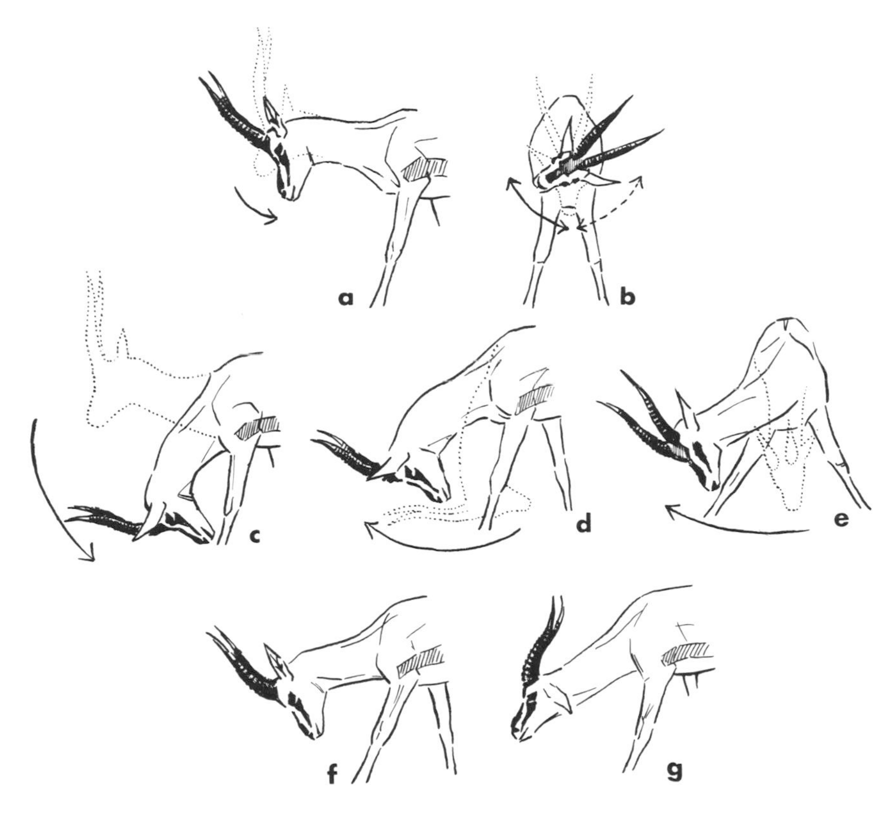

Figure 36: Common fighting movements in Antilopinae (a-f, adult male Thomson's gazelle; g, subadult Grant's gazelle). (a) Nod-butt. (b) Horn-levering (twisting). (c) Forward-downward blow. (d) Forward-swing. (e) Sideward-swing. (f) Posture of readiness. (g) Head-low posture.

(e) *Upward-push and upward-backward-push or-swing*—The upward-push is a nodding movement similar, or even identical, to the nod-butt, but the force of the push goes upward. When the neck is strongly involved in the movement, the upward-push may merge into an upward-backward-push or -swing in which the animal not only pushes upward but also somewhat sideward and, above all, backward, more or less in the direction of its own withers. In Antilopinae, this movement was only seen in springbok when fighting in parallel position (Walther 1981).

(f) *Sideward-push and sideward-swing*—The animal turns its head on the long axis so that both horns point to one side or the other. By a corresponding movement of the neck, the push goes in that respective direction. When the neck is strongly involved, the sideward-push becomes a sideward-swing (Figure 36e) in which the head and neck are moved in about the same way as the arms of a man in scything grain. Sideward-push and sideward-swing are particularly frequent in parallel fighting; however, they also can occur in a frontal fight. In the latter case, they serve as parrying maneuvers (e.g., in response to the opponent's attack by a forward-push or a downward-blow), and/or to tightly interlock the horns. In the beginning of a fight, some species may also use the movements under discussion to establish firm horn contact. Almost fantastic turns of the head are possible in such cases. For example, a Grant's buck may turn his nose toward the sky and then, "thread" his horns between those of the rival from above.

(g) *Horn-levering* (Figure 36b) *and horn-weaving*—In horn-levering, the head is held at shoulder level or lower, the chin is somewhat pulled toward the throat, and the animal twists its head alternately to the right or the left. Horn-weaving is an intensification of horn-levering. When the horns are turned to the right or left, the whole neck is swung in an arc in the corresponding direction. Thus, the movement is very similar to that in a sideward-swing but it goes alternately from one side to the other.

(h) *Head-low posture* (Figure 36g)—The animal lowers the head as in a grazing posture, but usually the nose is kept a few centimeters above the ground. The horn tips point upward or somewhat upward-backward. In this posture, the animal can easily catch the opponent's pushes and blows. Therefore, the head-low posture is frequently used for parrying the adversary's attack, but it is not exclusively limited to defense. For example, both opponents may display the head-low posture when initiating a fight. From this position, even a small aggressive movement, such as a nod-butt, may establish horn contact. Also, in reciprocal head-low posture, the rivals sometimes touch each other's noses or cheeks before they bring their horns together. One may speak of some kind of "taking measure" in such cases.

A number of types of fighting follow from these aggressive head and neck movements in combination with the leg and torso movements and according to differences in the intensity of the fighting actions. It is possible for a whole fight to be carried out exclusively according to one of these fighting types, but frequently several of them occur in one encounter.

(1) *Boxing*—Boxing occurs when nod-butting is not combined with any other aggressive movement, except perhaps a short rush or leap toward the opponent. In one-sided aggression, boxing may lead to body attacks. In reciprocal encounters, the horns of one fighter usually hit the other's horns. Boxing is quite frequent in females and young animals but infrequent in territorial males.

(2) *Horning*—This type contains all fighting movements in which no intensive neck movements are involved, and all are performed very gently and with low intensity. Horning is quite frequent in immature males, and also may occur in non-territorial adult males, but it has never been seen in territorial bucks.

(3) *Clash-fighting* (Figure 37b-d)—This is a very frequent type of fighting among territorial Antilopinae males. Reciprocal forward-downward blows or

push-butts, often immediately followed by forward-pushes, forward-swings, sideward-pushes, or sideward-swings lead to short, but violent horn contacts. After the clash, one or both combatants immediately leap back and frequently leap forward again into the next clash. The action of the legs is considerable. For example, in Thomson's gazelle, where clash-fighting is the most common type of fighting between territorial neighbors, the combatants jump toward each other with all four legs. They always spread the forelegs wide, and they sometimes kick into the air with the hindlegs at the moment of the clash. In jumping back, they close the forelegs again.

Figure 37: Border encounter between territorial blackbuck males. (a) Parallel march in nose-lifted display. (b) Turning into a frontal position and lunging together. (c) Clash. (d) Mutual turning aside after clash. (e) Males graze away from each other. (Photos: E.C. Mungall—Guajolote Ranch, Texas.)

(4) *Push-fighting*—This fighting technique usually does not occur on its own. Instead, it develops from clash-fighting or front-pressing (see below). The

most important fighting movements are the push-butt and the forward-push which sometimes are immediately followed by a sideward-push. In contrast to clash-fighting, the opponents do not jump back, and their horns remain in permanent contact. In contrast to front-pressing, the horns do not firmly interlock during push-fighting. Movements of the legs and torso result in vehement forward thrusts on the part of the attacker and in ducking down or retreat leaps (with the hindlegs) on the part of the defender. When his technique works, the defender may immediately come back with a forward thrust. One may consider push-fighting to be a transition between clash-fighting and front-pressing. In high-intensity fights between territorial Antilopinae males, push-fighting is seen fairly regularly.

(5) *Twist-fighting* (Figure 38)—The horns are crossed and more or less interlocked. The most important movements are horn-levering and horn-weaving. The one fighter forces the opponent into his rhythm of head-twisting which apparently is a rather convincing demonstration of his superior strength. Estes (1967) feels the purpose of twist-fighting is to break or sprain the opponent's neck. However, this never happened in the more than one thousand fights we observed in this study. Thus, the interpretation is somewhat unlikely. Twist-fighting occurs mainly in combination with, or as an interlude during, front-pressing.

(6) *Horn-pressing* (Figure 39)—The rivals approach each other with lowered heads and lean the frontal surfaces of their horns against each other. Then, each of them tries to press the horns of the opponent back toward the latter's nape. When the fight lasts somewhat longer, the horns almost invariably interlock, and horn-pressing merges into front-pressing. Thus, horn-pressing occurs mainly in the beginning of a fight. In Antilopinae, it seems to be more frequent in fights between non-territorial males than in encounters of territorial individuals.

(7) *Front-pressing (or forehead-pressing)* (Figure 40)—The rivals cross and firmly interlock their horns close to the bases. The foreheads of the fighters touch, or almost touch, each other. When the horns of the species are long enough, the head of each combatant is between the opponent's horns like a horse between the shafts of a cart. The tips of the horns point forward-upward or straight forward. The foreheads of the fighters are close to the ground or directly on it. Then, each tries with all his might to push the opponent back. Spreading the forelegs is obligatory in Antilopinae. Quite a number of the described fighting movements, such as forward-push and forward-swing, sideward-push, sideward-swing, head-low, horn-levering, and horn-weaving, may occur during front-pressing. The often considerable "leg-work" and torso movements may result in maneuvers which do not occur or are, at least, not as pronounced in other types of fighting. For example, both rivals may simultaneously wrestle and push against each other with powerful forward-thrusts. In other cases, one fighter may parry the other's forward-thrust by ducking deep (Figure 41). As mentioned above, the maneuver is suitable to let the attacker's forward-thrust run out. At the same time, it is the ideal position for the defender to come back with a powerful forward-thrust. The attacker may try to prevent such a counter-attack by stretching his back at maximum and by intensive forward stepping and pushing with his hindlegs (Figure 42). Provided that the defender is not

Figure 38: Twist-fighting in territorial springbok males. (a) In parallel position. (b) In frontal position. (Photos: F.R. Walther—Etosha National Park, South West Africa.)

Figure 39: Horn-pressing in Grant's gazelle (Photo: F.R. Walther—Serengeti National Park, Tanzania.)

Figure 40: Forehead-pressing in Grant's gazelle (non-territorial males). (Photo: F.R. Walther—Serengeti National Park, Tanzania.)

Figure 41: Forward-thrust (left) and parrying it by ducking deep (right) in a fight between springbok males. (Photo: F.R. Walther—Etosha National Park, South West Africa.)

considerably stronger than the attacker, he must immediately step or jump backwards in order to let this attack run out and to regain a firm stand for further defense or counter-attacks. Obviously, certain attacking maneuvers can only be parried by very definite defense maneuvers as in human boxing or fencing. Sometimes one of the combatants may even assume a defense posture, e.g., deep ducking and head-low, without being attacked and invite a definite form of the opponent's attack, e.g., a forward-thrust, in this way. Furthermore, in front-pressing, the firmly interlocked horns enable the rivals to use the sideward-throwing of the whole body as an enormously intensified form of horn-levering. Front-pressing is not lacking as a form of fighting in any of the Antilopinae species; however, there are differences in frequency. For example, it is very common in Grant's gazelle, whereas in Thomson's gazelle, it usually occurs only in very intensive and long fights.

Figure 42: Stretching the back in forward-thrust attack (left) and defense by jumping backwards (right) in fighting Thomson's gazelle.

(8) *Fight-circling* (Figure 43)—This occurs when the opponents have locked their horns so tightly that they temporarily have difficulties in breaking loose. In this situation, they pivot with their hindquarters while keeping their heads in the circle's center. Usually they can get their horns free during fight-circling, and then very frequently one of the opponents turns for flight. It is likely that fight circling may account for quite a number of broken horns.

(9) *Air-cushion-fighting* (Figure 44)—This type is not unusual in encounters between territorial bucks, and in Thomson's gazelle and mountain gazelle, it is even rather frequent. In air-cushion-fighting, the opponents perform attacking and parrying maneuvers as in clash-fighting, push-fighting, twist-fighting, and front-pressing, but without touching each other. It looks as if an invisible cushion had been placed between the horns of the combatants. Air-cushion-fighting may occur before or after a fight with horn contact or as an intermezzo during a fight with horn contact. Sometimes the whole encounter can be carried out by air-cushion-fighting, i.e., the combatants do not make any physical contact at all.

It is possible to distinguish several degrees of intensity in fights. Low intensity fights are fights of the horning type. Usually, they are exclusively carried out by head movements. Sometimes the neck may be somewhat involved. They hardly ever occur in territorial males. Considerable movement of the head and neck occur in fights of medium intensity. Sometimes leaps may be seen. The forelegs may be spread. The hindlegs may push forward moderately.

Figure 43: Fight-circling in Grant's gazelle (non-territorial males), moment of getting free. (Photo: F.R. Walther—Serengeti National Park, Tanzania.)

Figure 44: Air-cushion fighting between territorial neighbors in Thomson's gazelle. (Photo: F.R. Walther—Serengeti National Park, Tanzania.)

The torso is hardly involved in these actions. Naso-nasal contacts and "taking-measure" frequently initiate these encounters. The butts and blows with the horns are somewhat more violent than in low intensity fights. They usually are delivered from a head-low posture, and the horns of the opponents may cross each other. Medium intensity fights are infrequent in territorial males.

In high intensity fights, vehement aggressive movements of the head and the neck, intensive "leg-work" and often a strong involvement of the torso in the actions are typical. Clash-fighting, push-fighting, vehement twist-fighting, and front-pressing are characteristic of high intensity fights. Of course, the fighters move forward and backward according to their offensive and defensive maneuvers, but they usually do not displace each other over long distances. *Cum grano salis,* they remain on the spot. Most fights of territorial males are high intensity fights. In fights of the highest intensity, the attacker typically displaces the opponent over distances of 10 to 20 m by vehement push-fighting and/or front-pressing. Frequently, he also is displaced over a considerable distance in the very next moment. Sideward-throwing of the body and fight-circling occur only in fights of the highest intensity. This highest intensity is only reached in long and very severe fights and even then only temporarily. The fight may begin with high intensity, then reach highest intensity, and drop back to high intensity, and possibly reach highest intensity once more, etc.

Sometimes the combatants do not reach the same fighting intensity at the same time. The one may already be fighting with high intensity while his adversary still shows medium intensity. In such cases, the fighter who started out with the higher degree of intensity momentarily may appear to be superior; however, it is almost always the fighter starting with the lower intensity who eventually wins the fight. In a sense, he "warms up" after his opponent has exhausted his energies.

The duration of fights ranges from a fraction of a second (e.g., when the rivals clash together only once) up to half a day. However, fights of one or more hours are extremely rare. The majority range only from a few seconds to about five minutes. A 10 minute fight is already a long fight in Antilopinae. Except for one-clash encounters, a fight consists of several bouts interrupted by pauses during which the opponents simply stand in front of each other, show threat displays, mark objects, urinate and defecate, or graze, etc. Usually, the rivals remain more or less at the same place during such pauses, but sometimes they move ahead grazing parallel to each other, or walk in a parallel march. Exceptionally, they may even gallop side by side.

A fight can end with the flight of the defeated opponent. However, this ending is not as frequent as many people apparently assume. For example, in 741 fights of all sex, age, and social classes in Thomson's gazelle (Walther 1978a), only 16% ended with the flight of the defeated combatant. In fights among territorial neighbors, an end with the loser's flight is extremely exceptional. Slightly more frequently, the loser withdraws walking while the winner remains standing behind and possibly threatening toward the withdrawing adversary, or he may follow him walking in a pursuit march. Usually, however, there is no obvious winner or loser in the fights between territorial Antilopinae bucks. After they have fought for a while, one or both may abruptly switch to another activity (e.g., herding females), simply turn and walk off, or start grazing and move away from each other while grazing (Figure 37e). In Thomson's gazelle, the grazing ritual, i.e., both rivals grazing in a sequence of predictable position changes relative to each other, is even the usual ending of a fight among territorial neighbors (Figure 45).

Figure 45: Grazing ritual between territorial neighbors in Thomson's gazelle. (a) The combatants step backwards in frontal orientation after the last clash and start grazing. (b) They turn into a lateral position during grazing. (c) While grazing, they move parallel along the boundary. (d) They turn into a reverse orientation. Each of them grazes back to the center of his territory. (Photos: F.R. Walther—Serengeti National Park, Tanzania.)

When a combatant is unequivocally defeated in a fight, he may turn and flee, and the victor may chase him. As mentioned above, this is rare in fights between territorial individuals. On the other hand, territorial bucks frequently chase trespassing bachelors without fighting previously (Figure 46). The owner simply rushes toward the non-territorial male at a gallop uttering loud vocalizations (p. 99), and the bachelor turns for flight pursued by the territorial male until he has passed the boundary and left the territory. One may speak of a "symbolic" chase in such cases since the territorial buck treats the intruder as if he had defeated him in a fight. In a sense, he anticipates the victory. Under these aspects, "symbolic" chasing comes close to a threat display, i.e., an expressive behavior.

Figure 46: After a female herd intermingled with non-territorial males has entered the territory of a tommy buck, the owner (right) chases the bachelors out of his territory without any previous threat or fight. (Photo: F.R. Walther —Serengeti National Park, Tanzania.)

Threat and Dominance Displays

The term "dominance display" is used here in the sense of the term *"Imponiergebaren"* in German literature. Many authors do not distinguish between threat displays and dominance displays but refer to them both as threat displays. It is true that threat and dominance displays serve nearly the same functions, and that there are transitions between the two which sometimes may cause problems in distinguishing them. On the other hand, there simply is a difference between threat displays and dominance displays. Threat displays (in the strict sense) are much more closely related to fighting behavior than are dominance displays, i.e., threat displays either are precisely the same movements as in a fight or they are intention movements for fighting. In a dominance display, the sender tries to impress the opponent by demonstrating his size. The animal performs a movement or presents itself in a way which is suitable to intimidate or to challenge the opponent, but without showing an immediate readiness for fighting. Such dominance displays are particularly frequent in socially high-ranking individuals, and, thus, also in territorial males. However, some species have not evolved distinct dominance displays. Thomson's gazelle, for instance, has almost none.

Some threat displays are nothing but full-fledged fighting actions performed without touching the opponent and frequently at a distance from which they cannot possibly touch him. In our previous discussions, we mentioned object aggressions, air-cushion-fights, and "symbolic" chases (Figure 46). All these come, at the least, very close to such threat behaviors; one probably could even justify saying that they already are threats. Threats without any reservations are "fighting movements at a distance" such as "symbolic" nod-butt, head-low posture and horn-sweeping. In the last, the male abruptly lowers his head and horns to the ground, sometimes beating the grass and the ground with the frontal surfaces of his horns, straight forward or sidewards similarly as in a weaving action, and then raises his head as abruptly as he lowered it. Possibly, one could also include head shaking (as in the negation in humans) since it at least resembles the levering movements in twist-fighting. The direct approach toward the opponent is a somewhat problematic case. When it is combined with well-defined threat displays, such as medial or high presentation of horns (see below), there is no reason why it should not be classified as belonging to the threat displays in the strict sense. However, this direct approach also occurs in combination with dominance displays (e.g., erect posture) or quite frequently even without any additional display, and then one may be inclined to consider it more as being a dominance display.

With the exception of direct approach, air-cushion fight, object aggression, and "symbolic" chase, all threats which may be considered to be nothing else than full "fighting movements at a distance," are relatively infrequent in territorial Antilopinae males. Much more frequent are threat displays in which the animal performs only an intention movement for fighting, i.e., the very first, initial movement is "frozen" into a posture. In Antilopinae, such ritualized intention movements are predominantly the medial and the high presentation of horns (Figure 47). Head and horns are carried in a posture which obviously is the initiation of a butt or a blow. The head is at body level in medial presentation of horns and it is clearly above body level, i.e., the neck is more or less erect, in high presentation. There are great variations in the elaboration and frequency of these displays in the various Antilopinae species. For example, in Thomson's gazelle, medial and high presentation are frequent in adult males.

Figure 47: Medial (left) and high (right) presentation of horns in a boundary encounter of territorial tommy bucks. (Photo: F.R. Walther—Serengeti National Park, Tanzania.)

In territorial bucks, the high presentation of horns is even *the* prominent aggressive display. Also, in dorcas and goitered gazelles, medial and high presentation of horns are quite pronounced. Both displays are present in Grant's and mountain gazelles, but there are other aggressive displays (certain dominance displays) which are at least as important and frequent. High presentation of horns has never been seen in gerenuk and blackbuck. Medial presentation does occur, but, at least in blackbuck, other displays are more frequent and more striking. In springbok, high presentation of horns is lacking; the medial presentation is present although not very pronounced.

Springbok generally is the species with the least elaborate threat and dominance displays of all the Antilopinae. Besides medial horn presentation, one also may see "symbolic" nod-butting, head shaking, and head-low posture, the latter relatively most often. Interestingly enough, when presenting the horns or when in head-low posture, a springbok may perform small but vehement jerks with the head while simultaneously snapping with its mouth (Walther 1981) without biting or even touching the opponent. Such snapping movements are not reported from any other Antilopinae species.

As stated above, dominance displays are particularly typical of highly dominant individuals, and they have similar effects (intimidation or challenge) on recipients as threat displays, but without resembling fighting actions which involve use of the horns. At least some of the dominance displays may well be related to fighting techniques of hornless ancestors of our recent bovids (Walther, 1960a, 1979).

One of the principal dominance displays is the erect posture (which may also be involved as a component in the high presentation threat display). In the pure form of the erect posture (Figure 48), the neck is vertically erected and the long axis of the head points straight forward, i.e., the nose is carried somewhat higher than in a relaxed standing or moving posture. In species such as gerenuk, Grant's gazelle, and Indian blackbuck, the nose can be lifted higher to a nose-forward-upward-posture—or shorter: a "nose-lift"—when the display is intensified. In blackbuck, the display can even become a complete nose-up posture with the nose pointing nearly straight upward to the sky. However, this highest intensity is usually reached only in courtship, not in encounters with male rivals.

Further peculiarities of displaying blackbuck males are the ear-drop (i.e., the ears are turned downward so that their tips point toward the ground similar to the habitual ear posture of Indian cattle) and the raising of the tail which eventually is curled forward over the back. In Grant's gazelle, the tail can be raised to a horizontal position or somewhat above and may swing to the right and to the left. No special tail movements were observed or reported from other Antilopinae species in combination with erect agonistic displays. Likewise, there is nothing comparable to the ear-drop of blackbuck in the other species. However, when displaying from a more or less lateral position, the sender may "point" one ear toward the recipient.

The pure erect posture (with nose straight forward) and the nose-lift as well as the high presentation of horns can merge into a head-turned-away display, i.e., the head is turned about 45° away from the opponent (head-sideward inclination—Walther 1977b). Only in relatively rare cases is it turned as much as

90° in Antilopinae. This head-turned-away (Figure 49) can be displayed in frontal and in lateral orientation toward the recipient. Possibly with the exception of the springbok, it is present in all the Antilopinae species, but there are great differences in elaboration and frequency. In Grant's gazelle, the head-turned-away display is somewhat more pronounced and frequent than in others. However, it reaches its greatest elaboration in mountain gazelle where the head-turned-away and the high presentation of the horns are the two most striking and frequent displays of territorial males.

Figure 48: Erect posture in broadside position (right) in Grant's gazelle. (Photo: F.R. Walther—Serengeti National Park, Tanzania.)

Besides the erect posture and its modifications (lifting the nose, turning the head away, etc.), the lateral position toward the opponent (Figures 48, 49a) is the most common dominance display in Antilopinae, and frequently it is combined with an erect attitude. In a one-sided encounter (i.e., an encounter in which only one of the partners behaves aggressively), the sender may sometimes stand broadside in front of the recipient (lateral T-position). This blocking of the recipient's path probably is the most original form of the lateral display. More frequently in Antilopinae, the sender can also be in parallel or reverse-parallel (head-to-tail) position with a recipient, especially in a reciprocal encounter (both partners displaying). Parallel orientation of the two adversaries usually leads to a parallel march (Figure 37a). As with the head-turned-away, the parallel march probably is not completely absent in any Antilopinae species, but it occurs with different frequencies. For example, in Thomson's gazelle, it is rare or even exceptional. In Grant's and mountain gazelles, it is considerably more frequent. The parallel march is most common and pronounced in blackbuck. When the rivals are in reverse-parallel position, they may remain standing or may start circling each other, which is most pronounced and frequent in Grant's gazelle.

Figure 49: Head-turned-away display. (a) Reciprocal display during circling in reverse-parallel position in Grant's gazelle. (Photo: F.R. Walther—Serengeti National Park, Tanzania.) (b) Head-turned-away display (left) in frontal orientation during an encounter in territorial male mountain gazelle. (Photo: G.A. Grau—northern Negev.)

On the whole, Grant's gazelle is the Antilopinae species in which the dominance displays have reached their greatest elaboration. The sender positions himself broadside to the recipient, be it in lateral T-position, parallel or, most commonly, reverse-parallel orientation. He erects his neck or even bends it somewhat back toward his withers. He lifts his nose forward-upward and he turns his head in a more or less pronounced fashion away from the opponent. Then, he vehemently turns his nose toward his opponent. After this head-flagging (Figure 50), he returns to the erect nose-forward-upward posture and may repeat the head-flagging if the recipient does not walk off immediately. In a reciprocal encounter, both opponents may repeat the head-flagging simultaneously or alternately. This head-flagging display of Grant's gazelle is unique among the Antilopinae species.

Figure 50: Head-flagging of a territorial Grant's gazelle buck (right) in repelling non-territorial intruders (left and center background). The subordinate recipient (left) shows self-grooming induced by the head-low posture in response to the territorial male's display. (Photo: F.R. Walther—Serengeti National Park, Tanzania.)

Fights very infrequently follow directly after dominance displays. After the peers have displayed to each other for some time, but neither has given in, they usually change to threat displays which may be followed by a fight. Thus, threat displays are inserted between dominance displays and fights. In such a transitional stage of an agonistic encounter, it can also happen that threat components occur in combination with components of dominance displays. The most common of these threat-dominance displays is sideward angling of horns. The animal stands or moves in erect posture and in lateral position (dominance display components) and tilts its horns (threat component) sidewards toward the opponent.

Space-Claim Displays and Excitement Activities

Before, after, or during pauses in agonistic encounters, certain movements may be seen which are not aggressive behaviors, or, at least, are not as clearly aggressive as are threat and dominance displays. These movements occur even more frequently in non-agonistic situations. They often are termed "displacement activities" in ethological literature. We wish to avoid this term for several reasons which are discussed more in detail elsewhere (Walther 1974, 1979).

Some of these movements are directed toward the ground or the vegetation, and it seems that they have something to do with the space around the animal. "Translated" into human words, they may mean: "This is *my* place!" When they are used in agonistic encounters, one may perhaps speak of space-claim displays. They rarely occur in the agonistic encounters of non-territorial, adult Antilopinae males, and they never occur in the encounters of females or immature males. They are more or less restricted to interactions between territorial neighbors where space-claims meaningfully fit the situation, even though their intimidating or challenging effects upon the recipients often are not as strong as those of threat or dominance displays. In Antilopinae, such space-claim displays are predominantly the marking activities discussed above: scraping the ground, urination and defecation, object aggression, and marking with preorbital glands.

The grazing ritual of Thomson's gazelle also may be included in the space-claim displays. Grazing in connection with agonistic encounters is quite common in bovids, and it obviously has somewhat different meanings in the different species. Generally, it seems to be a transitional action (Lind 1959) induced by lowering the head. However, lowering the head in an agonistic encounter can have different meanings in bovids. A fighter can lower his head for an aggressive attack. In this case, it is an offensive movement. However, a fighter can also lower his head (head-low posture—see p. 104) in order to parry an opponent's attack. In this case, it is a defensive maneuver. When used as an expressive display and in response to an offensive threat display or a dominance display by the challenger, the defensive head-low posture may even express a certain inferiority. Moreover, lowering the head in an agonistic encounter can also be submissive behavior (see p. 121). According to these different meanings of lowering the head, the meanings of grazing can vary correspondingly. Ritualization processes probably have focused the meaning of agonistic grazing more toward submission in some of the species and more toward offensive aggression in others. When watching how easily territorial tommy males switch from fighting to grazing and back to fighting, one may come to the conclusion that their grazing does not have much to do with submission but is more on the aggressive side. Furthermore, distance and orientation components are involved in the agonistic grazing of territorial tommy bucks which justify speaking of it as a grazing ritual in this species (Figure 45).

To understand this ritual, one must know that (a) aggressiveness is greatest in all these males when they meet each other frontally, (b) generally a gazelle has the tendency to turn its head somewhat sideways while grazing, and (c) in non-territorial animals the average distance which they can keep between individuals varies with sex and activity. The distance is greater between males than between females, and it is greater in grazing than in any other activity (Walther 1977a).

Of course, there are variations in these grazing rituals, but the typical case can be described as follows. Two territorial tommy bucks meet at the boundary of neighboring territories. They threaten and fight each other. After several clashes, they simultaneously begin to graze and step backwards for 1 to 2 m. This close range as well as the frontal orientation and the grazing posture (lowered head) enable them to immediately fight again, if necessary. If they do not renew the fight, permanent grazing brings about further enlargement of the distance to 2 to 5 m which corresponds to the average distance between non-territorial males during grazing. The rivals also turn more and more into a parallel or reverse-parallel position which is a common orientation among grazing gazelles and from which renewed attacks are considerably less likely than from the frontal position, although they may occasionally occur. Furthermore, the combatants mutually seem to block each other's paths and to prevent each other from proceeding into the neighbor's territory by this reciprocal broadside position. The longer this phase lasts, the greater the distance becomes. In lengthy parallel grazing along the territorial boundary, the rivals may easily separate 10 to 20 m or even more from each other. Sooner or later, they turn into a reverse (hindquarters to hindquarters) position. After both combatants have done so, a renewal of the fight is very exceptional. Their orientation and the general tendency to move slowly ahead during grazing carry them farther and farther away from each other. Each of them grazes back toward the center of his territory. Thus, there is no winner and no loser in these encounters which apparently serve to establish and later, to ratify the position of the territorial boundaries among peers. As mentioned above, agonistic grazing may be more related to a defensive attitude or even to submission in other Antilopinae species. However, in Thomson's gazelle and in blackbuck, the described grazing ritual definitely shows features which relate it to the space-claim displays, e.g., that it is the most common ending of agonistic encounters among territorial bucks, whereas it rarely occurs in non-territorial adult males and is lacking in females and subadult males.

Other non-aggressive behaviors which may occur in connection with agonistic encounters, are lacking any relation to space, do not contribute anything to the solution of the encounter, and do not have any effects upon the partners except occasional contagion (an animal taking up performance of the same behavior as is being performed by another animal close by). These behaviors may be called "excitement movements" (Walther 1974) because, beside their normal everyday functions, they may occur in any situation in which the animal is somewhat more excited than usual (e.g., when watching a predator at a distance, before crossing unfavorable terrain, etc.), and thus, also during hostile encounters. One may consider them to be outlets of inner tensions. Such excitement activities are predominantly self-grooming with hooves or mouth, shaking the body, stepping in place with all four legs, stepping around in a small circle, leaping in place, and, very exceptionally in Antilopinae, alert watching and uttering vocalizations which resemble "alarm"-calls. Such excitement activities in connection with agonistic encounters are considerably less frequent in Antilopinae than in some other bovid species such as wildebeest (Estes 1969) or kongoni (Gosling 1974). In territorial individuals, only self-grooming plays a certain, although minor role.

Submissive Displays

It appears advisable to discuss the submissive displays in connection with aggressive displays because they are more or less the opposite of dominance and offensive threat displays, although at least some of them seem to be related to defensive threat displays. In offensive threat displays, the animal assumes a posture and an orientation toward the opponent from which it can immediately attack. Submissive postures are inadequate for an attack, and the horns are more or less turned away from the opponent. In dominance displays, the animal tries to appear as tall and/or broad as possible. During submissive displays, it tries to appear as small and thin as possible. In threat and dominance displays, the sender is frontally or laterally oriented toward the recipient. During extreme submission, he orients with his hindquarters toward the recipient. Thus, the most extreme form of submission possible in these animals is lying down with head and neck stretched forward on the ground and with the hindquarters toward the opponent. In the Antilopinae investigated, this extreme form of submission was only exhibited by captive animals (probably due to spatial limitations), but not by free-ranging individuals except when females not fully in heat were relentlessly courted and sexually driven by a male. Even in this situation, lying down by the female is infrequent in the wild. The most common submissive display of Antilopinae is lowering the head as an intention movement for lying down. In rare cases, the submissive animal may even drop down to its "knees," i.e., carpal joints (Figure 51). The submissive lowering of the head differs from the aggressive or, at least, defensive head-low posture in that a submissive animal stretches its head and neck more forward so that the nose is more or less parallel to the ground. On the other hand, this submissive display certainly is similar to the defensive head-low, and the animal may switch from one posture to the other. Both submissive display and defensive head-low can induce grazing, and it is very possible that the submissive lowering of the head has originated from the defensive head-low posture, i.e., from those cases in which the latter is used in response to an offensive threat or a dominance display of the challenger. Quite frequently, the submissive animal orients with its back toward the dominant, but this is not necessarily so. Even frontal orientation is not unusual. This submissive lowering of the head and neck can be displayed in standing, walking (Figure 52), and even while fleeing in full gallop. A special tail movement occurs only in the gerenuk. The submissive animal lifts and curls its tail over the back (Walther 1961, Leuthold 1971).

As long as a territorial male is on his own ground, he does not show any submissive behavior. However, even a territorial male may display features of submission when he has temporarily left his territory, particularly when he must cross other territories on his way to or from his ground. In the species investigated, this situation occurs most frequently in blackbuck. A typical report (Cary 1976b), which at the same time exemplifies the differences in the behavior of the same male inside and outside of his territory, may illustrate such a case (Figure 53).

When a territorial blackbuck was returning after an absence of several hours, he took a route that cut diagonally across a neighboring territory. When he noticed the neighbor grazing some distance away in his territory, the returning buck began to run. At this moment, the owner approached him in the nose-lifted display along an intersecting course. Now, the returning buck, still holding to his original route, lowered his head and neck forward into the full submissive posture as he ran. Barely getting past, he sped across his border and dramatically changed to the nose-lifted posture himself, even giving a loud series of "róho" calls as he settled into a slow, pacing gait. His pursuer, who had given up the chase at the border, let him go and resumed grazing.

Figure 51: Adolescent gerenuk male (left) drops down on its "knees" with head and neck stretched forward and with tail curled in submission when threatened by the object aggression of an adult buck. (Photo: F.R. Walther—Frankfurt Zoo, Frankfurt a. M. Germany.)

Functions and Situational Motivations of Aggressive Interactions

In Antilopinae as well as bovids in general, aggressive interactions occur in all sex, age, and social classes. Thus, aggression serves a great variety of social functions in these animals, and the situational motivations vary accordingly. Only the territorial encounters are of interest in this discussion, but there are many other situational motivations of aggression in these animals, and there is hardly any part of their social life in which aggression is not involved (Walther 1978a).

In territorial Antilopinae males, aggressive behavior serves or, at least, essentially contributes (a) to establish territorial boundaries, and later (b) to ratify their position among territorial neighbors as well as (c) to prevent other, territorial or, more often, non-territorial males from invading the territory or, in special cases (see p. 176), to dominate them and keep them in a subordinate status as long as they are inside the territory. These three are the most general, frequent, and important functions and situational motivations of territorial aggression in Antilopinae. However, there are several more, although less frequent or of minor importance. In the special case when bachelors are occasionally tolerated within a territory, (d) maintaining or enlarging the individual distance may sometimes motivate aggressive behavior of the territorial buck toward the bachelors. Furthermore, territorial aggression may sometimes result in (e) usually minor changes of the position of the boundary between neigh-

Figure 52: Adolescent blackbuck male (right) assumes submissive attitude in withdrawing from a territorial male in dominance display. (Photo: E.C. Mungall—Eagle Ranch, Texas.)

bors. Another case (f) is the acquisition of an occupied territory by conquest and defense against it. Conquest attempts happen occasionally, but they are definitely rare in Antilopinae. For example, former bachelors tried to conquer an occupied territory only three times during the three years of study of the Thomson's gazelle, and only once a bachelor succeeded in conquering a territory (see p. 180) in the two and a half year study of blackbuck. Furthermore, (g) a territorial buck may behave aggressively in herding females. In most cases, he chases them from the boundary toward the center of his territory. At least in some of the species under discussion, these chases are quite frequent. Finally, (h) components with aggressive overtones are involved in the male's courtship behavior, and when a female is extremely uncooperative, a courting male may become directly aggressive toward her. According to these functions, the most common and general aggression releasing situation for a territorial buck is: "(another) male entering the territory."

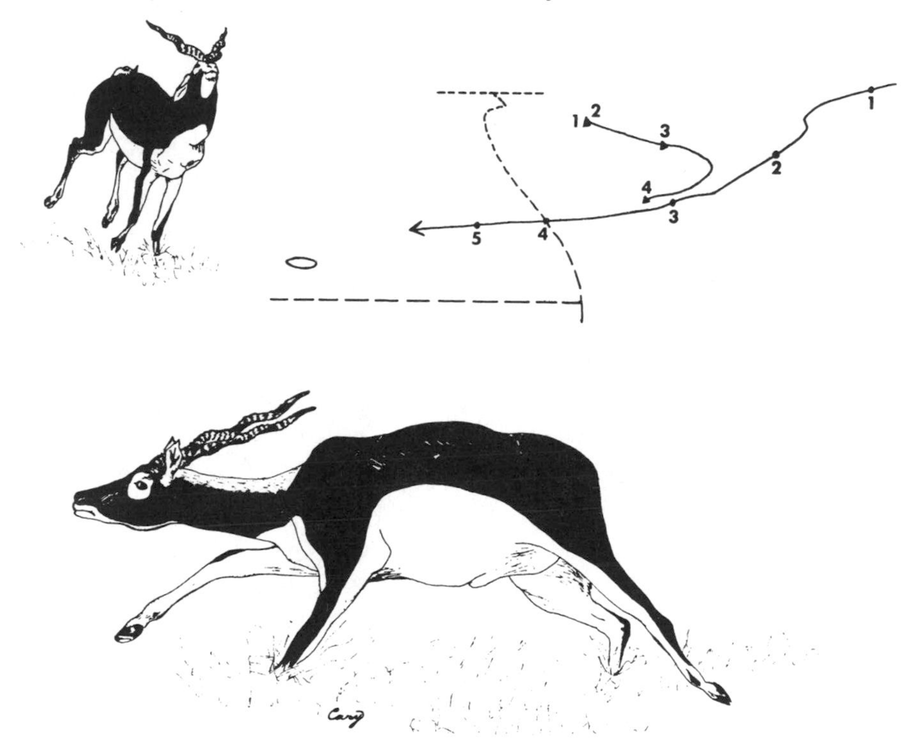

Figure 53: Two adult blackbuck males in an encounter involving both nose-up display (buck left background represented by triangles) and submissive display (buck center foreground represented by dots). Dashed lines are territory boundaries. Oval is a dung pile in "Dot's" territory. (For details of the situation see text.) Time 1: "Triangle" is grazing as "Dot" comes into view. Time 2: "Dot" starts to run as "Triangle" stops grazing and moves toward him. Time 3: "Dot" assumes extreme submissive posture as "Triangle" assumes nose-lifted posture and starts running. Time 4: "Dot" slows and assumes nose-lifted posture as he crosses into his own territory. "Triangle" ceases pursuit. Time 5: "Dot" vocalizes and then returns to normal posture.

When establishing a territory, the male "informs" other males of his territorial status and the location of the occupied area by his aggressive actions. Thus, frequency and severity of such aggression largely depend on the number of other males in the vicinity and other circumstantial factors. Most frequent are the interactions when the territorial hopeful was a member of a bachelor group, and he now tries to establish his territory within the formerly mutual home range. In this situation, the non-territorial males very frequently enter the territory, and it apparently takes them quite a while to learn that they are now excluded from a definite part of their home range. Very severe interactions may occur when two males simultaneously establish neighboring territories. In such cases, it may occasionally even happen that one of them is totally defeated in a fight, flees at a full gallop while being chased by the victor far beyond the realm of the latter's territory, and loses his territorial status before he has fully established it. Also, when a buck returns to a formerly held territory after a long absence and another buck has established his territory there meanwhile, the hostilities between the new and the old owner can be unusually intense and frequent over a rather long period of time.

Boundary ratification encounters take place between neighbors after the territories are well established. Sometimes they remain display encounters. More often the peers fight each other, however, there is neither a winner nor a loser in such fights. Their frequency largely depends on size and density of the territories. In species such as gerenuk, mountain gazelle, and Grant's gazelle, encounters between territorial neighbors are infrequent because the territories are large and often there are broad unoccupied areas between them. In species with smaller, adjacent territories, such as Indian blackbuck, boundary ratification encounters are more common. When many relatively small territories are together in a dense mosaic, as is frequent in Thomson's gazelle, each territorial buck has a minimum of two to three boundary ratification encounters per day. A considerable amount of spontaneity seems to be involved in such cases. For example, a territorial tommy buck frequently walks to the boundary without being challenged in any form, stops there and waits until the neighbor arrives, so that they can threaten each other and have a fight right on the boundary. Obviously, these boundary ratification encounters of species with small, adjacent territories contribute to maintaining the territorial status of the males. In a sense, a territorial male without neighbors and without the obligatory daily encounters with them, is no longer a territorial buck. When all territorial males except one (or a very few widely distributed over the formerly densely occupied field) leave a given area, this lonely male becomes insecure about the position of the boundaries and sooner or later he gives up and leaves his territory.

Territorial invasion by territorial neighbors, i.e., that one of them clearly trespasses the boundary and goes far into the other's territory, is rather rare in Antilopinae, except in blackbuck, in which territorial males frequently leave their territories for brief periods for feeding or drinking and then have to cross other territories on their way from and back "home." As another example, in Grant's gazelle, under certain conditions (see p. 176), a territorial buck may leave his territory together with a mixed herd which has been with him for a while, may temporarily move together with them and in this way enter the territory of another buck. Relatively most frequently, a territorial gazelle buck invades a neighboring territory while intently herding or chasing a female. In such a situation, he sometimes simply may be unaware of having crossed the boundary. Usually, however, such a buck stops right at the boundary, even from a full gallop, and does not trespass.

Non-territorial males frequently invade territories. These invasions do not appear to be voluntary violations of the boundary on the part of the bachelors. They simply happen to trespass more or less by chance. When charged by the owner of the territory, they immediately withdraw or flee in the majority of the cases.

Aggressions by territorial bucks toward non-territorial males to maintain or enlarge individual distance, occur only in the relatively rare cases (see p. 184) when a territorial buck temporarily tolerates bachelors in his territory. He is then moderately dominant over them, displays to them, chases or (seldom) fights them when they happen to cross his path.

Aggression (chasing) by a territorial buck toward females due to territorial establishment, is restricted to one special and rare situation: when a male becomes territorial and has extraordinarily many encounters with other males, he may occasionally become so aggressive for a few hours that he chases away any conspecific which happens to enter his territory, regardless of sex.

Somewhat more frequent are cases in which chasing a female is directly related to sexual behavior. When a female is intensively courted by a territorial buck but she does not respond to his courtship, he may "lose patience" and chase her at a gallop. She then may leave his territory. Often the owner of the next territory has watched the chasing action of his neighbor and will now take over and chase this female himself. In this way, the chase may be continued through several neighboring territories. Finally, the female is exhausted and responds to the courtship of the last male. Among the species investigated, these sexual chases were recorded with some frequency only in Thomson's gazelle and blackbuck. Even here, however, the majority of the mating rituals proceeded without such chases.

Chasing is more common during herding the females in a territory. In this case, the territorial buck is not chasing the females away but is trying to move them more toward the center of his territory. Gently butting the female with the horns is a regular part of the male's courtship behavior only in mountain gazelle. He touches the female's hip, flank or shoulder with the anterior or lateral surface of his horns.

Interactions with Opponents of Different Classes

It seems to be a widespread rule in ungulate behavior that the animals interact with conspecifics of their own sex, age, and social class most frequently, next in frequency with animals of neighboring classes, but infrequently or not at all with animals from more distant classes (Tables 5-7). Correspondingly, one could expect that territorial males will have most of their agonistic encounters with each other, fewer with adult non-territorial males as the next neighboring class, still fewer with immature males, very few with females and almost none with juveniles. This expectation is correct here in that territorial males hardly ever become aggressive toward juveniles and that they have considerably fewer agonistic interactions with immature males than with adults. Also, their behavior toward females would fit the general rule if one were to ex-

Table 5: Distribution of Agonistic Encounters on Sex, Age, and Social Classes (in Percentage) in Thomson's Gazelle

versus	te ♂	ad ♂	sa ♂	ao ♂	ad ♀	im ♀	hgf	faw + nfa	N Encounters
te ♂	28.1	26.9	4.6	3.8	31.7	4.8	0.1		1,726
ad ♂	1.9	74.6	15.8	2.3	4.8	0.6			1,630
sa ♂		24.2	51.1	24.2	0.5				182
ao ♂		4.4	17.6	60.2	14.4	0.6	2.8		181
ad ♀	6.4	0.5	0.5	15.1	43.7	10.5	5.9	17.4	219
im ♀	5.1				7.7	53.9	12.8	20.5	39
hgf				7.8	32.8	6.3	43.7	9.4	64
faw + nfa					52.3	2.3	3.9	41.5	128

Abbreviations:
te ♂ = territorial male; ad ♂ = non-territorial adult male; sa ♂ = subadult male; ao ♂ = adolescent male; ad ♀ = adult female; im ♀ = immature (adolescent + subadult) female; hgf = half-grown fawn; faw = fawn; nfa = neonate fawn.

One-sided encounters are listed only in the class of the animal which behaved aggressively. Reciprocal encounters between combatants from different classes are listed twice, i.e., in each class of the two opponents. The percentages are summed to 100 horizontally.

Table 6: Distribution of Agonistic Encounters on Sex, Age, and Social Classes (in Percentage) in Grant's Gazelle

versus	te ♂	ad ♂	sa ♂	ao ♂	ad ♀	im ♀	hgf	faw + nfa	N Encounters
te ♂	4.6	43.1	15.0	6.8	22.1	6.2	1.3	0.9	453
ad ♂	8.5	74.2	14.6	2.2	0.5				417
sa ♂		13.8	76.2	8.5	1.5				130
ao ♂			15.4	66.2	13.8		4.6		65
ad ♀	32.5		1.9	13.0	35.2	5.8	1.9	9.7	154
im ♀	19.0				38.1	38.1		4.8	21
hgf				11.8	41.1		35.3	11.8	17
faw + nfa					52.4		9.5	38.1	21

Abbreviations and remarks: See Table 5.

Table 7: Distribution of Initiation of Agonistic Encounters on Sex, Age, and Social Classes (in Percentage) in Mountain Gazelle

	. . . . Addressee				
Addressor	te ♂	ad ♂	im ♂	ad ♀	N Encounters
te ♂	7.8	40.1	20.7	31.4	232
ad ♂	2.9	55.3	25.9	15.9	170
im ♂	8.7	17.4	63.7	10.2	69
ad ♀			24.3	75.7	37

Abbreviations:
te ♂ = territorial male; ad ♂ = non-territorial adult male; im ♂ = immature (adolescent + subadult) male; ad ♀ = adult female

The addressor is the animal which initiated the encounter.

The percentages are summed to 100 horizontally.

clude chasing while herding. However, chasing during herding is, at the least, a behavior with aggressive overtones. Thus, it can hardly be excluded from a discussion on aggression. Consequently, aggressive interactions of territorial males with females usually turn out to be more frequent than expected. Also, the frequencies of agonistic encounters with each other and with non-territorial adult males often somewhat deviate from the general rule. In the best case, the numbers of interactions with territorial neighbors and those with adult bachelors equal each other, but usually the encounters with non-territorial adult males are more frequent than those among territorial males. This result is not surprising when one takes into account how easily and often the numerous bachelors enter territories. Furthermore, the number of interactions among territorial neighbors becomes small with large and/or more distant territories, as discussed above (p. 125).

Territorial Male Versus Territorial Male (Figures 54a, 55a)—Except for the rare cases in which a bachelor male tries to conquer an occupied territory, only another territorial male is an equivalent opponent to the owner of a territory. One-sided displays are rare between territorial bucks. If a species possesses several threat and/or dominance displays, they may occur in a sequence, the dominance displays always before the threat displays. Also, space-claim displays are quite frequent, e.g., in Thomson's gazelle, 36.7% of 485 territorial boundary encounters were carried out exclusively by grazing rituals. Depending on species-specific peculiarities, roughly 50 to 70% of the encounters among territorial males can be settled by displays. However, the challenge by reciprocal dominance and threat displays leads to fighting in a considerable portion of such interactions. For example, in Thomson's gazelle, fights occurred in 53.6% of 485 interactions between territorial bucks. In Grant's gazelle, the proportion of fights was 42.9% in 21 encounters, and it was 28% of 18 encounters in mountain gazelle (Grau 1974).

Figure 54: Forms of aggression and decisions (or carrying out) of agonistic encounters in territorial males of Thomson's and Grant's gazelles with opponents of different classes. Abbreviations: te♂ = territorial male; ad♂ = non-territorial adult male; sa♂ = subadult male; ao♂ = adolescent male; ad♀ = adult female; vs. = versus. Each form of aggression is counted only once per encounter. Thus Figure 54 shows in how many encounters a given form of aggression was observed. It does not show how often this behavior pattern was performed. The formulation "or carrying out" (of agonistic encounters) refers to the relatively few cases without a clear resolution. Figure 54 presents forms of aggression which did not decide an encounter but were followed by another form of aggression or occurred after the decision had been made (open block diagrams), and forms of aggression which decided the encounters (cross hatched block diagram). Reference figures for both are the numbers of encounters, but only the percentages of the decisions add together to 100%. In this way, it is possible to figure the effectiveness of the single forms of aggression from these histograms. For example, one-sided horn threats occurred in 7% of the (485) encounters of territorial male versus territorial male in Thomson's gazelle, and none of them decided an encounter; reciprocal horn threats occurred in 49.1% of these (485) encounters, and 2.5% of the (485) encounters were decided by them, etc. (For more details on data collection, principles of evaluation, and tests of significance see Walther 1978a).

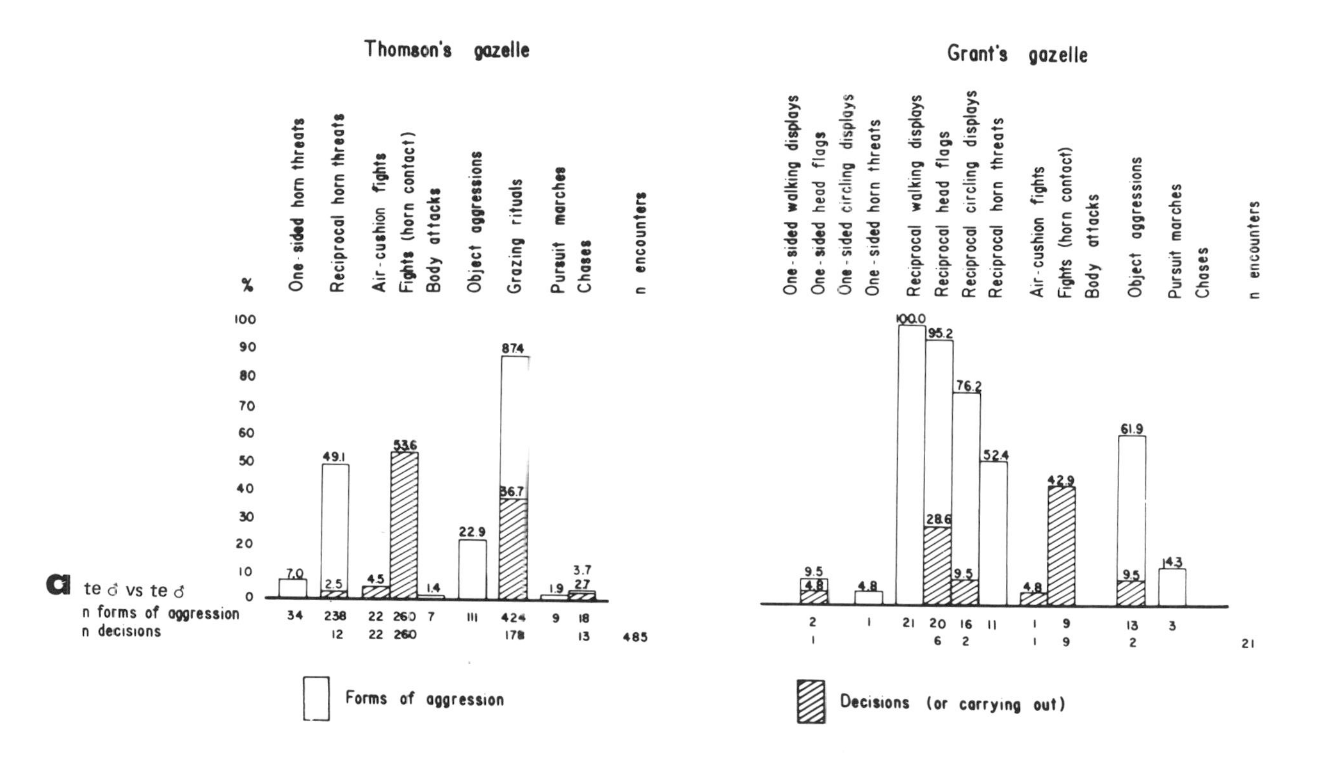

Figure 54: (Legend on page 128)

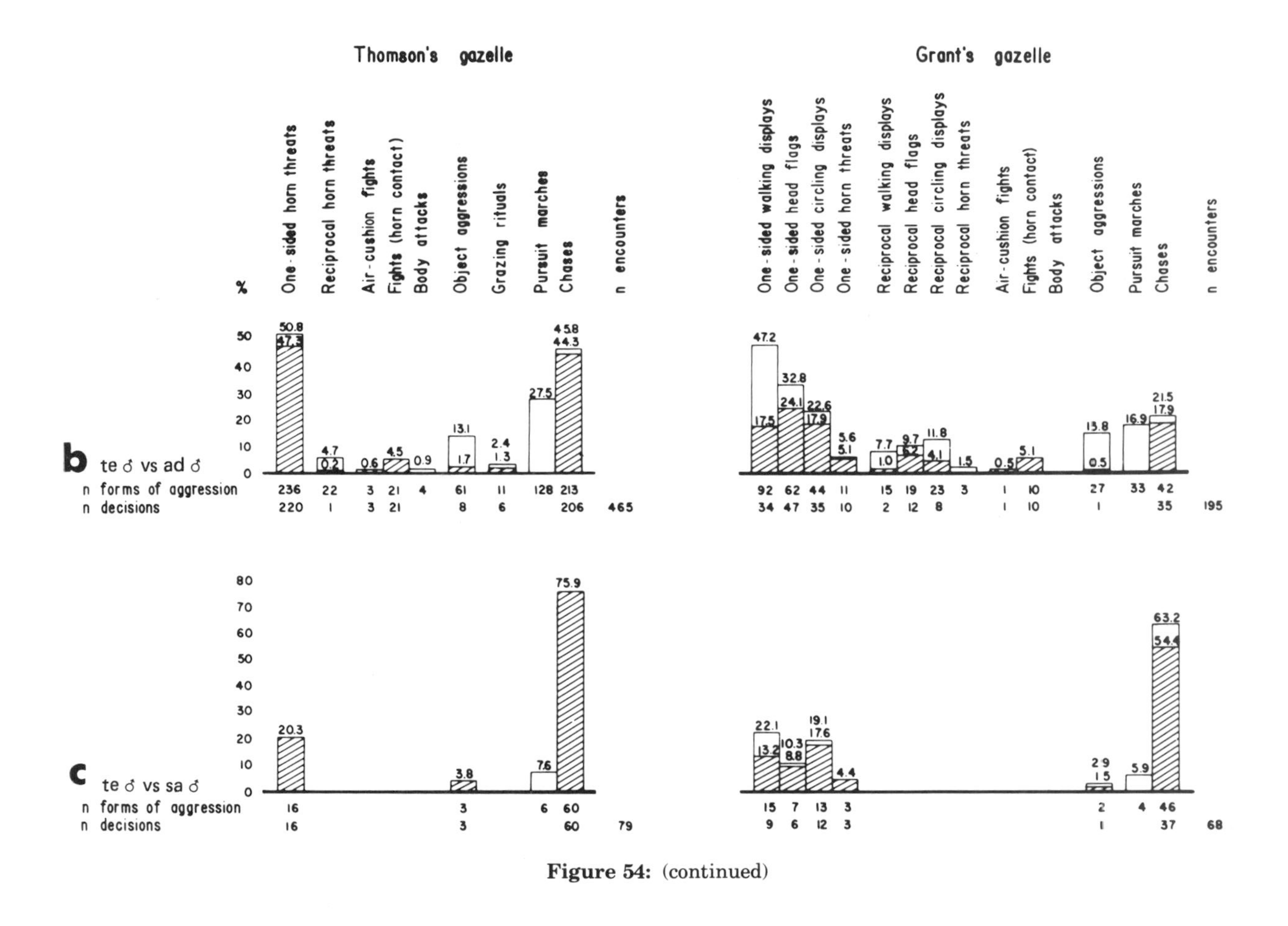

Figure 54: (continued)

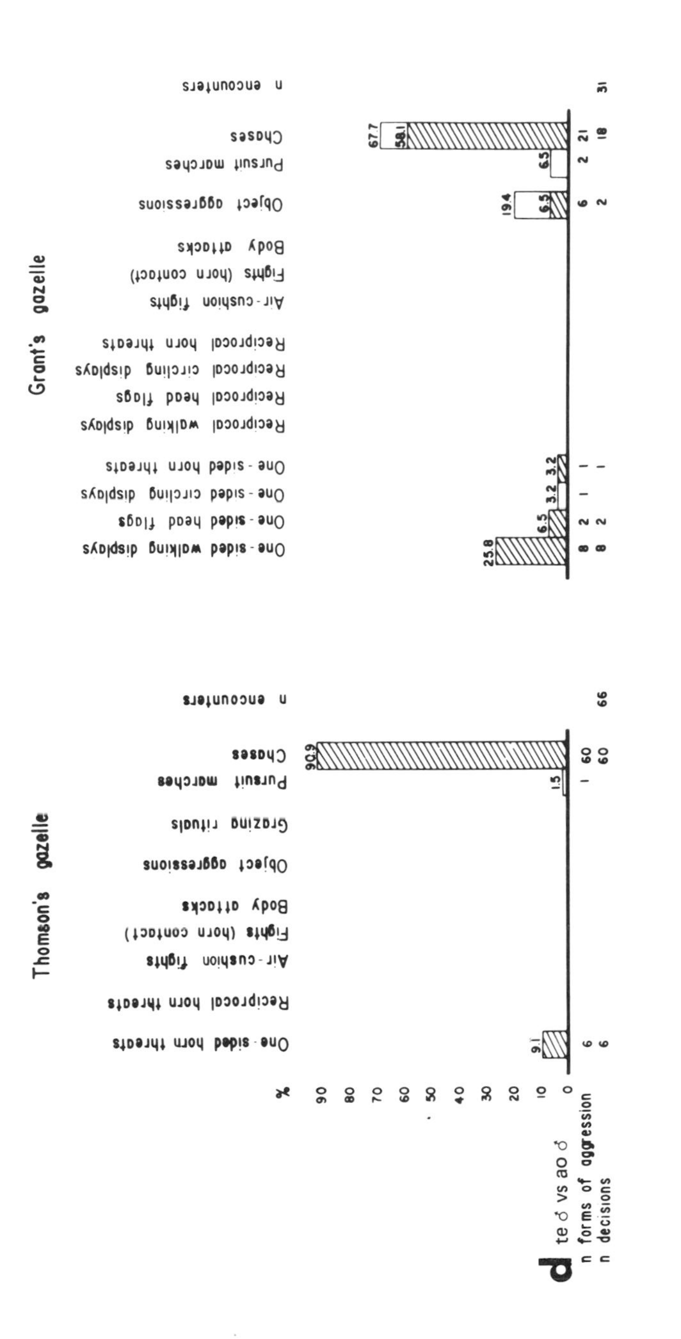

Figure 54: (continued)

other activities such as herding females, etc. Withdrawals of one of the combatants with the other following him in a pursuit march or flights with the victor chasing the fleeing opponent at a gallop, are infrequent, e.g., in Thomson's gazelle, withdrawals and flights together occurred in only 5.6% of 485 encounters.

Territorial Male Versus Non-Territorial Adult Male (Figures 54b, 55b)—As stated above, interactions of territorial bucks with adult bachelors are at least as frequent, but often considerably more frequent, than those with territorial neighbors. For example, in 1726 agonistic interactions of Thomson's gazelle (Table 5) in which at least one of the combatants was a territorial buck, 28.1% were among territorial neighbors and 26.9% with non-territorial adult bucks. In Grant's gazelle (Table 6), only 4.6% of 453 encounters involving territorial individuals were between territorial bucks, but 43.0% with adult bachelors. In mountain gazelle, 8% of 232 agonistic encounters were between territorial males, but 40% were with non-territorial adult males (Table 7). In blackbuck, 18.6% of 1295 interactions were between territorial males, and 26.3% were with bachelors.

Cases in which an adult bachelor challenges and even fights the owner of a territory, are rare. Somewhat more frequent are the cases in which a territorial buck charges an adult bachelor and the latter threatens and fights back. In short, encounters of territorial with non-territorial adult bucks which end in a fight, do occur, but they make up only a minor portion of the total. For example, out of 465 agonistic interactions of territorial Thomson's gazelle males with non-territorial adult males, only 6.6% ended with fights; or in Grant's gazelle, 18.9% of 195 encounters; in mountain gazelle, 4% out of 93 encounters; in blackbuck 12.4% of 298 encounters. Thus, an adult bachelor sometimes can be an equivalent opponent to a territorial male, but more often he is not.

Correspondingly, one-sided encounters are frequent in interactions of territorial bucks with adult bachelors, i.e., only the territorial male behaves aggressively, and the bachelor withdraws or even flees without showing any signs of aggression. In one-sided encounters with non-territorial adult males, the territorial bucks may use threat and/or dominance displays, sometimes also space-claim displays (particularly object aggression), frequently followed by a pursuit march, as well as chases without any previous display. Already the one-sided displays, but even more so the pursuit marches and the chases indicate that the territorial male does not treat the bachelor as an equivalent opponent because the pursuit marches are rare and chases are even exceptional in the encounters among territorial peers. The number of one-sided (threat and/or dominance) displays may approximately equal or exceed the number of chases without prior displays. For example, one-sided threats settled 47.3% (plus 1.7% cases of one-sided object aggressions), and chases without prior displays settled 44.3% of the 465 encounters of territorial tommy bucks with adult bachelors. In Grant's gazelle with their more elaborate dominance displays, 65.1% of the corresponding 195 encounters were decided by one-sided displays of the territorial bucks and 17.9% by chases. Of 298 corresponding blackbuck encounters, 47% were decided by one-sided displaying and 26.5% were settled by chases. One-sided displays by territorial mountain gazelle ended 97% of 93 encounters with non-territorial adult males.

Territorial Male Versus Immature Male (Figures 54c and d, 55c)—Under natural conditions, agonistic encounters of territorial males with immature males are considerably less frequent than encounters with adult bachelors for two reasons. First, the subadult and adolescent classes comprise a much smaller life span than the adult class. Hence, the numbers of immature males are considerably smaller than those of adult males in a natural population. The possibility of encounters of territorial bucks with adult bachelors is consequently much greater than that with immature males. Secondly, when bachelors of different age classes together invade a territory, the owner usually charges the adult bachelors first. Of course, the young males become aware of the encounters of the territorial buck with their adult companions, and they frequently leave the territory before the owner focuses on them.

Consequently, encounters with subadult males made up only 4.6% and those with adolescent males 3.8% (as compared to 26.9% with adult bachelors) in all the 1726 encounters involving territorial Thomson's gazelle bucks (Table 5). In Grant's gazelle (Table 6), the corresponding proportions were 15% with subadult males and 6.8% with adolescent males (as compared to 43% with non-territorial adult bucks) in 453 encounters. In mountain gazelle (Table 7), out of 232 agonistic interactions involving territorial males, 21% were with immature males (as compared to 40% with adult bachelors).

Reciprocal displays, fights, and grazing rituals are absent in the encounters of territorial bucks with immature males. Even pursuit marches are infrequent. The territorial males rely completely on one-sided displays and chases without prior displays; however, in contrast to encounters with adult bachelors, the chases outnumber the threat and dominance displays. When a distinction between adolescent and subadult males was possible, even a difference in the treatment of these two classes of immature males became evident. The territorial bucks chase the adolescent males proportionally more often than the subadult males. For example, out of 68 encounters of territorial Grant's bucks with subadult males, 44% were settled by one-sided threat and dominance displays (plus 1.5% by one-sided object aggressions) and 54.4% by chases. In 31 encounters with adolescent males, 35.5% were decided by one-sided threat and dominance displays (plus 6.5% by one-sided object aggressions) and 58.1% by chases. Even more striking is the relative preponderance of chasing in Thomson's gazelle which do not have dominance displays, and are restricted to the use of high and medial presentation of the horns. In this species, 20.3% of 79 encounters of territorial bucks with subadult males were decided by one-sided threats (plus 3.8% by object aggressions), and chases were used in 75.9% of the cases. However, in the 66 encounters of territorial tommy bucks with adolescent males, only 9.1% were settled by one-sided threats while 90.9% were settled by chases without prior displays.

In short, the immature males are hardly treated as opponents by the territorial bucks; more like an "irritation" which they try to get out of their territories, like people chasing little children or pets out of a flower-bed.

Aggression Toward Females (Figures 54e, 55d)—While agonistic interactions of territorial males with males of their own class and with bachelors of different ages largely appear to follow the same principal lines in the different Antilopinae species, aggression toward females shows more species-specific differences. However, with the inclusion of chasing of females during herding, the aggression of territorial males toward females has one point in common in all the species. It is considerably more frequent than aggression of non-territorial adult and subadult males toward females (which was found to range from 0 to 15% in the different species), and it makes up quite a proportion in the agonistic encounters of the territorial bucks. For example, in Thomson's gazelle (Table 5), 36.5% of the 1726 agonistic encounters involving territorial bucks were with females, 28.3% of 435 encounters in Grant's gazelle (Table 6), 31% of 233 encounters in mountain gazelle, and 43.4% of 1295 interactions in blackbuck.

The simplest case is in Thomson's gazelle because (a) tommy females neither attack adult males nor defend themselves when charged by an adult male, and (b) the adult tommy bucks have an almost completely different behavioral repertoire toward other males than toward females. Object aggression and horn threats by territorial tommy males toward females are so exceptional that they can be ignored under quantitative considerations, and with very little exaggeration, one may generally say that adult tommy bucks do not show aggressive behavior toward females. The one very remarkable exception is the chase. Of 547 aggressive acts by territorial tommy bucks toward adult females, 98.7% were chases. Thus, the increasing frequency of chases by territorial tommy bucks toward conspecifics of different classes (2.7% in the encounters among territorial neighbors, 44.3% in encounters with non-territorial adult males, 75.9% with subadult males, 90.9% with adolescent males) is continued and culminates in their aggression toward females. Clearly, a female is hopelessly inferior to an adult buck in this species.

As far as our present knowledge goes, the situation probably is more or less the same in several other gazelle species such as dorcas gazelle, goitered gazelle, and gerenuk. In mountain gazelle, it is at least similar in that about 80% of 73 interactions of territorial males with females were chases. However, threat and dominance displays and occasionally even short fights occurred in these encounters with small but not quite insignificant frequencies. Moreover, the male's courtship includes ritualized butting of the female with his horns in this species.

In Grant's gazelle, only 37% of the 100 observed aggressive interactions of territorial bucks with females were chases. 33% of these encounters were settled by one-sided threat and dominance displays and 18% by one-sided object aggression by the bucks. 6% of the encounters were decided by reciprocal displays and another 6% by fights between territorial males and females. This picture is not much different from that of aggressive interactions of territorial bucks with non-territorial adult males. Thus, in Grant's gazelle, females are also inferior opponents to territorial bucks, but they certainly are not as hopelessly inferior to them as female Thomson's gazelle are to conspecific males. This view is strongly supported when one takes into account that in the 68 observed encounters of territorial Grant's bucks with subadult males as well as in the 31 encounters with adolescent males, reciprocal displays and fights were lacking, and the proportions of interactions decided by chases without prior displays made up 54.4% and 58.1%, i.e., the proportions of chases were considerably greater than in the encounters with adult females in this species. At least in part, this may be explained by my (Walther's) impression that Grant's females are generally somewhat more aggressive than the females of most other Antilopinae species.

The situation is particularly complicated in blackbuck. On the one hand, the blackbuck females do not have horns whereas the adult males have horns of considerable length. Consequently, there are no fights between territorial bucks and females, and the females are hopelessly inferior to adult males. On the other hand, a territorial blackbuck male may sometimes use his horns in herding females, and above all, he frequently uses the same dominance display (erect nose-forward-upward posture, ear-drop, erecting and curling the tail over the back) in aggressive encounters with opponents as in herding and courtship toward females. Under these aspects, one might say that the behavioral repertoire of the blackbuck male is the same toward males and females, and that he treats males and females alike. Moreover, there are numerous cases in which it is hard to decide whether the male's approach toward females is "meant to be " aggressive or sexual.

HERDING BEHAVIOR

When females have entered a territory or have approached its boundary very closely, the male frequently will drive them toward the central region of his territory (Figure 56). When the females later are going to leave, the male will almost regularly try to keep at least some of them in his territory. For these behaviors, the term "herding" is commonly used in the literature. However, this territorial herding is a very special kind of herding. For one, it is restricted to females, males are not herded in Antilopinae, and the herding buck always goes after only one female at a time, even when several try to leave the territory simultaneously. If the male succeeds in driving one female in a definite direction, the others may follow her. Alternatively, he may have to herd every single one of them. Secondly, this territorial herding is not analogous to rounding-up in Antilopinae, i.e., the buck does not herd the females together into a dense bunch. On the contrary, when members of a female herd are widely dispersed within a territory, the male does not herd them together. Territorial herding only occurs when a female is approaching the boundary, and it serves to bring her toward the center of the territory. Although courtship and herding can be related to each other in several regards, and although herding can sometimes change into courtship, the territorial herding should be distinguished from sexual behavior because it is independent of the female's estrous state. A territorial Antilopinae male always herds the females in his territory, regardless whether he can court them or copulate with them later.

It is imperative that the male places himself between a female and the boundary and blocks her path when preventing her from leaving the territory. Therefore, when he is positioned behind her, he first has to pass her. Even in the most intensive form of herding, when chasing a female at full gallop (Figure 57), the buck tries to pass her and to come into a position in front of her. Of course, this can only be recognized after he has reached her. As long as the female is running in front of him and/or when she is faster and reaches and crosses the boundary before he has passed her, a herding chase cannot be distinguished from an aggressive or sexual pursuit.

Figure 56: Territorial herding. (a) A tommy female (left) has left the territory of a buck (grazing in background left) and entered the territory of his neighbor (center). (b) The latter drives her with nose-lifted display toward the center of his territory. The other male (right) remains grazing in his territory. (Photos: F.R. Walther—Serengeti National Park, Tanzania.)

Figure 57: Chasing the female in territorial herding. The tommy buck is passing the female in order to block her path. (Photo: F.R. Walther—Serengeti National Park, Tanzania.)

When the buck is able to pass the female or when he has been positioned between her and the boundary from the beginning, he may block her path in lateral T-position and keep her at the spot for some time. Sooner or later he orients frontally toward her and drives her back at a walk. In this situation, the male may perform object aggressions or, in some species, other threat displays such as medial presentation of horns or, in blackbuck, horn-sweeping. However, dominance displays are more common than threat displays in the herding actions. In blackbuck, the erect posture with lifted nose, ear-drop and curled tail is very typical. In Grant's gazelle, the buck also displays the erect posture with lifted nose, sometimes a head-turned-away or, still from lateral position, a sideward tilting of horns, and occasionally even head-flagging. However, "courtship" displays are the most common displays of Antilopinae males in territorial herding.

The term "courtship display" is somewhat misleading in this case, but at least it is brief. More precisely, these are displays performed predominantly by adult males in encounters with females. Thus, they frequently can be seen when a male is courting a female; however, they are not restricted to courtship. In species such as blackbuck, dominance displays (predominantly given toward other males) and courtship displays are much the same. Also in Grant's gazelle, the erect posture occurs as a dominance display in encounters between male opponents, although it is never as persistently displayed in agonistic interactions as it is in sexual encounters. The most striking use of courtship displays for herding occurs in species such as Thomson's gazelle in which the males have a different behavioral repertoire in their encounters with females from that used with other males. Here, head-and-neck stretched-forward posture, nose-forward-upward posture, nose-up, and the "drum-roll" (short alternating fore-leg kicks) are the major displays of territorial bucks when herding females. As previously mentioned, the males may also utter vocalizations in intensive herding, which usually are the same as in aggressive chases of bachelors such as "pshorre - pshorre - pshorre" in Thomson's gazelle, "róho - róho - róho" in blackbuck, "ru - ru - ru" and "ru-oo" in springbok, etc.

In Grant's gazelle—particularly in those cases in which a relatively stable group of females (harem) remains together with the same territorial buck for weeks and months—the herding actions described so far, only occur when the females are right at the boundary and are determined in their efforts to leave the territory. However, a territorial Grant's buck almost constantly signals to the females in which direction they should go or should not go while the herd is moving inside his large territory. Chasing, vocalizations, threat, dominance, and courtship displays do not occur in this "silent herding." It is merely accomplished by the male's position relative to the females when he is executing his usual maintenance activities such as moving, grazing, standing, lying. For example, in grazing, standing or resting, he may place himself in lateral T-position in front of the foremost female of the herd and in this way prevent her from advancing in a given direction. He also may graze or walk frontally toward her and make her turn, or he may follow behind her in grazing or walking and "push" her ahead, or move in front of her, releasing her following reaction and "leading" her in a given direction, etc. (Walther 1977b). As long as the females do not deliberately try to leave the territory, this "silent herding" works very well, but it is rather inconspicuous to the human observer. Probably for this reason, it has only been described in Grant's gazelle up to now, but it also is used by blackbuck and presumably exists in more species.

SEXUAL BEHAVIOR

First Contact and Testing Phase

Several phases are distinguishable in the mating rituals of Antilopinae, but they are not equally pronounced in all species. They also may vary within a species because one phase may pass into the next without a clear-cut caesura, an advanced ritual may drop back to an earlier phase for some reason and then the corresponding phases are repeated, or a male in great sexual excitement may start out immediately with the last phase, but the female may not yet be receptive and then he has to go back to an earlier phase. Despite such variations and special problems, however, the phases are recognizable, and so it seems advisable to discuss the mating rituals according to phases.

When females arrive and stay in a territory for some time, due to the male's herding actions, the buck begins to "test" them. He approaches one of them and, performing species-specific courtship displays (see below), he drives her for a short distance. Sometimes, she may flee at a gallop and may even leave his territory. However, usually she responds to his approach by walking in front of him, and she urinates after a short time. The male may or may not let the female's urine flow over his nose. In Antilopinae, however, he regularly sniffs at the female's urine on the ground and performs *Flehmen* (Figure 58). He raises his neck, usually keeping his head on a horizontal plane or occasionally lifting his nose above horizontal, he opens his mouth and remains motionless in this posture for several seconds. Sometimes he may turn his head somewhat to the side. Antilopinae males may somewhat retract the upper lip in *Flehmen,* but they do not curl it as do some other bovids. They frequently end *Flehmen* by licking their upper lip while nodding with the head.

Figure 58: *Flehmen* in several gazelle species. (Note that the males do not curl the upper lip.) (a) Thomson's gazelle. (b) Grant's gazelle. (Photos: F.R. Walther—Serengeti National Park, Tanzania.) (c) Mountain gazelle. (Photo: G.A. Grau—Research Zoo, Tel Aviv University.) (d) Sömmering's gazelle. (Photo: F.R. Walther—Hannover Zoo, Hannover, Germany.)

It is presumed (Backhaus 1961, Knappe 1964, Estes 1972) that the male can learn something about the female's estrous condition by performing *Flehmen.* (It should be mentioned that *Flehmen* occasionally occurs in other situations and even in females—particularly frequently in blackbuck—and also that smelling substances other than urine can release it.) In any case, the male either loses all interest or continues and reinforces his courtship after *Flehmen.* When he continues it, the female may urinate again, and the whole procedure may be repeated several times. When the male discontinues his courtship after *Flehmen,* he may approach the next female and possibly "test" all the females in his territory, one after the other. Certainly, the female's urinating and the male's *Fleh-*

men mean an important caesura in the mating ceremony, and many rituals end at this early stage.

In most Antilopinae species, males do not frequently touch the genitals or other regions of the females' bodies with their noses. Only the mountain gazelle male more or less regularly investigates the female's genital, inguinal, or hock region while standing behind her. In gerenuk, the male may mark the female with his preorbital glands (Figure 34).

Of course, a presupposition of successful courtship is that the female is on her feet. However, a female sometimes may be resting when the male starts courting her, or she may lie down as a kind of submissive behavior in response to the male's approach. In such cases, the buck has to make the female stand up. Quite frequently, the male's approach is sufficient. If the female does not rise so readily, he may push her with his nose or gently butt her with his horns. However, the Antilopinae buck most frequently uses his foreleg, i.e., he either paws her back or touches the female's rump with a *Laufschlag*-like movement or displays true *Laufschlag* (the latter predominantly in springbok). The female may not react at first or may even show defensive threat movements (e.g., winding the neck in gerenuk, thrusting neck and head upwards in blackbuck, etc.), but sooner or later she rises.

Demonstrative Driving Phase; Male Courtship Displays

When the courtship ritual is continued after *Flehmen,* a phase follows which is characterized by intensive displays by the male. He already may have shown these displays when soliciting urine from the female, but he now repeats and emphasizes them. If the female is not receptive, she may flee and possibly leave the male's territory. A Grant's gazelle female occasionally shows counter-displays more or less equivalent to those of the buck, and may even have a short fight with the male. In other Antilopinae species, a female occasionally may, at the very most, gently butt the male's body with her horns or forehead, display symbolic nod-butting, or lift her nose in a defensive gesture. In most of the cases, the female does not show any defense or displays comparable to those of the male. Quite frequently, she submissively lowers her head, and then her head-low posture may induce grazing. Grant's gazelle females also stretch their tails horizontally. However, the female's most common and typical response to the male's courtship displays in all the species is to walk in front of him in a moderately ritualized withdrawal. Obviously, it is the function of the male's courtship displays to make the female move ahead in Antilopinae—and therefore, these displays can also be used in situations other than courtship.

Except for some minor species-specific peculiarities—such as ear-drop in blackbuck, the sideward spreading of the ears (in the beginning of courtship) in dorcas gazelle and springbok, the horizontal stretching of the male's tail in Grant's gazelle and goitered gazelle, or the raising and forward-curling of the tail in blackbuck and springbok—the male courtship displays are essentially the same in all Antilopinae species. Species-specific differences mainly are brought about by different combinations as well as differences in presence, elaboration, and frequency of the single displays.

The most important courtship displays of Antilopinae males are: the head-

and-neck-stretched-forward posture (or the neck-stretch), the erect posture, the nose-forward-upward posture, and the foreleg-kick (or *Laufschlag*).

The neck-stretch (Figure 59) is very elaborated in Thomson's gazelle where it is the first display which a male regularly shows in any encounter with a female. It is also quite pronounced in dorcas gazelle and in goitered gazelle. In springbok, it is somewhat diminished, i.e., the male does not stretch his neck as stiffly and horizontally forward as e.g., a tommy buck. Forward stretching of the neck can also be seen in mountain gazelle and Grant's gazelle; however, here it is much less obligatory and it is often performed in such an accidental manner that one may doubt whether it is truly a special display in this species. The neck-stretch is lacking in gerenuk and in Indian blackbuck.

Figure 59: Neck-stretch of the courting tommy buck with widely opened preorbital glands. (Photo: F.R. Walther—Serengeti National Park, Tanzania.)

The erect posture (Figure 60) occurs in three forms of intensification. First, the male erects his neck more or less vertically (in blackbuck, the neck is somewhat more flexed in a "J-shaped" posture than in the other species) with the nose pointing horizontally forward. Sometimes a head-turned-away display may be combined with this posture. Probably, this erect posture is not completely lacking in any of the Antilopinae species. However, it is a pronounced display only in goitered gazelle, dama gazelle (Mungall 1980), Sömmering's gazelle, and possibly dibatag. In a number of other species such as blackbuck, Grant's gazelle, Thomson's gazelle, gerenuk, and mountain gazelle, it usually occurs only as a transitional movement merging into a nose-lift.

The nose-lift (Figure 61) is an erect posture in which the animal raises its nose up to about 45° above the horizontal level. Apparently, it is lacking in Sömmering's gazelle and is only poorly developed in dama gazelle. It was infrequently observed in dorcas gazelle. In goitered gazelle and springbok, it is definitely present but in a somewhat diminished form as compared to other species. In Thomson's gazelle, it is a well-pronounced movement which regularly follows after the neck-stretch when the male continues his approach toward the female. It is also present in gerenuk, mountain gazelle, and possibly dibatag. Finally, in Grant's gazelle and Indian blackbuck, the nose-lift is the most prominent display of the courting male and is held almost continuously throughout the entire ritual. In Thomson's gazelle, Grant's gazelle, gerenuk and, to a lesser extent, blackbuck, the nose-lift can be combined with foreleg-kicks in which the foreleg is not lifted more than about 45°. In some species such as Grant's gazelle, this *Laufschlag* in combination with the nose-lift usually takes the form of a big, stiff-legged step, in others, such as Thomson's gazelle, that of a drum-roll, i.e., the buck alternately kicks both forelegs, and in species such as blackbuck, it commonly is a few rapid, stiff-legged, tripping steps with the forelegs.

Figure 60: Erect posture of the courting male in dama gazelle. (Photo: E.C. Mungall—San Antonio Zoo, Texas.)

The nose-up posture (Figure 62) with the nose pointing almost straight up to the sky is the most intensified form of the erect posture. It only occurs in a few species, and is only displayed for a few seconds at a time. Grant's gazelle perform the nose-up imperfectly by jerking the nose somewhat higher than in the usual nose-upward-forward posture but without erecting it vertically. In Thomson's gazelle, the nose-up is well-pronounced but displayed relatively infrequently. The highly pronounced nose-up is most frequent in Indian blackbuck. It has not been observed in all the other species under discussion. As with the nose-lift, the nose-up can sometimes be combined with *Laufschlag;* in Grant's gazelle, this combination is even the rule.

Figure 61: Nose-lift display in courting males. (a) Blackbuck. (Photo: E.C. Mungall—Y.O. Ranch, Texas.) (b) Thomson's gazelle. (Photo: F.R. Walther—Serengeti National Park, Tanzania.)

The foreleg kick (Figure 63) occurs in two forms in Antilopinae. In the first form, the leg is raised to an angle of about 45° and frequently, although not necessarily, is combined with a nose-lift or a nose-up as described above. In the other form, the foreleg is raised to nearly 90°, and usually is executed in normal, or in an only slightly erect, posture with the nose pointing forward. Only in dorcas gazelle, may *Laufschlag* sometimes be combined with a neck-stretch. The 45°-*Laufschlag* usually does not touch the female. The 90°-*Laufschlag* may go between the female's hindlegs or laterally touch her hindleg from outside. However, even the 90°-*Laufschlag* can be performed without touching the female and sometimes is performed at a distance from which it cannot possibly touch her. The latter is particularly frequent in springbok. As with other courtship displays, the male can stand in any orientation with respect to the female when executing *Laufschlag;* however, most frequently he is standing or walking behind her. In some species such as gerenuk, foreleg-kicks can occur throughout the entire mating ritual, while in other species, they are restricted to definite parts of the ritual. For example, in Thomson's gazelle, the drum-roll with nose-lift is used in the beginning, whereas the 90°-*Laufschlag* in normal posture is typical of an advanced mating ritual after a phase more or less without foreleg-kicks.

Figure 62: Nose-up display in blackbuck. (Photo: E.C. Mungall—Y.O. Ranch, Texas.)

Males may sometimes use their horns when courting females in mountain gazelle, blackbuck and apparently also dama gazelle. In blackbuck, the male may thrash his horns down over the female's haunches when driving her ahead. Mountain gazelle males usually touch the female's hip gently with the anterior surface of their horns. Both species use the horns most frequently during advanced phases of the mating ritual.

Mating March and Pre-Mounting Phase

The vivid displays of the male during the demonstrative driving phase sooner or later result in the female walking in front of the male. Male and female circling in reverse-parallel orientation (mating-whirl-around) is not an obligatory part of the mating ritual in Antilopinae; however, it occurs relatively often when the female is somewhat, but not fully, in mating condition. In this situation, she may temporarily try to evade the male's driving actions by an abrupt turn of 180°. When the male continues his driving, they circle around each other in head-to-tail position. If the female is very persistent in evasive circling, the male may eventually cease his courtship.

Figure 63: Several forms of the foreleg-kick in different species. (a) 45°-*Laufschlag* combined with nose-lift in Grant's gazelle. This is the only form of the foreleg-kick in this species. (It is rainy season. The buck has mud on his hooves and withers.) (b) 90°-*Laufschlag* in normal posture by a tommy buck. The form of the foreleg-kick shown in (a) for Grant's gazelle also occurs in this species. (Photos: F.R. Walther—Serengeti National Park, Tanzania.) (c) Bent-kneed performance combined with touching the female's back with the male's snout in dama gazelle. Such "lazy" performances of the foreleg-kick occasionally occur in all the species. The dama buck also uses the 45° stiff-legged form of *Laufschlag,* and he does not necessarily touch the female with his snout. (Photo: E.C. Mungall—San Antonio Zoo, Texas.) (d) *Laufschlag* between the female's hindlegs in gerenuk. This form of the foreleg-kick also occurs in other species, and it is not the only form in gerenuk either. (Photo: F.R. Walther—Frankfurt Zoo, Frankfurt a. M., Germany.)

In driving the female forward, the male ceases or, at least, diminishes his displays after a while. The demonstrative driving phase merges into the phase of the mating march (Figure 64) in which the female walks forward uninterruptedly and the male follows closely behind her in tandem. The male's displays are diminished to different extents in the different species under discussion. For example, in Thomson's gazelle, Sömmering's gazelle, and gerenuk, the male frequently stops all courtship displays completely and follows the female while walking in normal posture, whereas in Grant's gazelle and Indian blackbuck, the male usually keeps the nose-lifted attitude although not necessarily as pronounced as in the beginning. Consequently, in some of the Antilopinae species, the phase of the mating march can be better distinguished from the demonstrative driving phase than in others. Blackbuck pairs frequently walk in big arcs, whereas in other species, the mating march proceeds along a zigzag course. When a herd of females is in a territory, the mating march often carries the pair away from the group and toward the peripheral regions of the territory.

Figure 64: Mating march. (a) Thomson's gazelle. (b) Sömmering's gazelle. (Photos: F.R. Walther—Serengeti National Park, Tanzania, and Hannover Zoo, Hannover, Germany.)

The mating march can be followed by a pre-copulatory phase which again varies with species in frequency of occurrence. In this phase, the female frequently makes pauses in walking during which she stands or grazes. Mountain gazelle and springbok females even stand most of the time. The males respond by an increase of the courtship displays which, in some species such as Indian blackbuck and Grant's gazelle, are the same as in the driving phase (pronounced nose-lift, nose-up, with or without foreleg-kick, etc.). In other species, they are different, e.g., the tommy buck uses the high *Laufschlag* in normal posture but not the neck-stretch, the nose-lift or the nose-up which are so typical of the demonstrative driving phase in the beginning of the ritual in this species. Mountain gazelle and blackbuck males frequently touch the female with their horns.

From the female's tendency for standing and the male's reinforced driving actions, a push-driving develops in which the female stops walking, the male intensively displays, the female walks ahead, stops again, the male displays, etc. During this push-driving, the female may again perform some aggressively tinged movements of defense but without directing them toward the male. Standing with her hindquarters toward the male, she may now threaten toward the space in front of her. If another female or a young animal happens to cross her path in this situation, she may direct her displays toward it or even butt it. However, such defensive behaviors, except the head-low posture, are infrequent in this phase. Only in Grant's females, a movement similar to head-flagging is quite common. The Grant's bucks almost regularly respond to it by mounting.

Mounting Phase and Copulation

Single mounting attempts may sometimes occur already during the phases described above. However, after the mating march or after push-driving, the male comes to a phase in which one complete mount follows another in a series. This series of mounts eventually ends with copulation. Four mounts before copulation appear to be the minimum number. Ten to 20 mounts are quite normal, and 30 to 40 mounts before copulation are by no means unusual. Occasionally, the number can be considerably larger, such as 164 mounts before copulation observed in one mating ritual of mountain gazelle.

The female carries head and neck erect during the mounting phase in Antilopinae. She stands or walks immediately before copulation. Apparently, there are minor differences among the single species, e.g., in mountain gazelle and springbok, the female usually stands prior to the male's mounting, whereas tommy females walk continuously in at least 50% of the cases. Even when the female has been standing before mounting, she starts walking at the moment the male mounts in all the species. The buck follows bipedally behind the female. Thus, one may say that mounting and even copulation can take place during walking in Antilopinae.

In mounting (Figure 65), the Antilopinae males carry head and neck in erect posture, and sometimes, particularly in blackbuck, they may lift their noses above horizontal. The male's forelegs more or less bend toward his chest, an attitude which strikingly differs from the stiff-legged stretching of the leg in

Figure 65: Sexual mounting in several gazelle species. (a) Dama gazelle. (Photo: E.C. Mungall—San Antonio Zoo, Texas.) (b) Gerenuk. (c) Springbok. (d) Thomson's gazelle. (e) Mounting . . . (f) . . . and copulation in Grant's gazelle. (Photos: F.R. Walther—Frankfurt Zoo, Frankfurt a. M., Germany, Etosha National Park, South West Africa, Serengeti National Park, Tanzania.)

a well-pronounced *Laufschlag*. The male does not cling to the female's hips with his forelegs. Sometimes, he may touch the female's croup with his chest. However, such cases appear to be "mistakes" in Antilopinae, in that the male mounted too hastily, or did not correctly "calculate" his position or distance relative to the female. In a "good" performance, a mounting Antilopinae male erects his body so steeply that he does not touch the female with his chest. As a matter of fact, in a perfect mount, he does not touch her at all except with his genital region. Thus, the mounting posture of an Antilopinae male is already very erect, but it becomes even more erect in copulation. It looks as if the buck might fall over backward at the very next moment (Figure 65f). In Thomson's gazelle, this literally happened on three occasions.

After copulation, the mating ritual abruptly ceases in Antilopinae. The animals do not copulate more than once during a ritual. After a pause of ordinarily at least half an hour, the male again may court the female, provided that she has remained in his territory, and may copulate with her again. Whether with the same or another female, no more than six copulations per day were observed by any Antilopinae male during the studies under discussion, and each copulation took place after a new ritual. When a male successfully courts females several times on the same day, he sometimes proceeds through the initial phases faster than in the first ritual, or he may even skip them completely. When a male has been courting a female and the ritual is prematurely interrupted, he may later start a mating ritual with another female beginning more or less with that phase in which the former ritual had ended. In such a case, his success largely depends on the female's readiness for copulation. Such tentative sexual encounters seem even to be the rule in mountain gazelle. Therefore, it is not easy to make statements about the time needed for a complete mating ritual in this species.

Also, in the other species, the length of a complete ritual often is not easy to determine due to interruptions and pauses, and because, under field conditions, one cannot often be sure of noticing the male's very first approach. Forty-five Thomson's gazelle rituals were observed from the male's very first "testing" of the female to copulation. The shortest ritual (after the female had been chased by other males through several territories) took only about 3 minutes. Twenty-two of these rituals were completed within 5 to 15 minutes, 13 within 15 to 30 minutes, and two within 45 to 60 minutes. Only one ritual (out of 45) took more than one hour (about 90 minutes). The 31 recorded mating rituals in Grant's gazelle show a similar time distribution. However, there were a few rituals which lasted considerably longer, and the longest took approximately 10 hours. Also in blackbuck, most of the (6) rituals observed from the first beginning to the end lasted 3 to 40 minutes. Thus, most of the Antilopinae courtship rituals that are continued without remarkable pauses are completed within 5 to 45 minutes. Copulation itself takes but a few seconds.

Postcopulatory Behavior

After copulation, the mating ritual comes to an abrupt end in Antilopinae. The male's sudden return to a relaxed "normal" posture is particularly striking. There is no conspicuous and obligatory postcopulatory behavior in Antilopinae. The male usually stands for a moment behind the standing female. The blackbuck female stands briefly with back arched and tail raised, then resumes a normal posture. In Grant's gazelle, the female frequently turns her head through 180° and glances back at him. Then, both male and female lose interest in each other and return to their usual maintenance activities. One or both partners may pace away, start grazing, or bed down for a rest. Sometimes the buck may groom his genital region but this is by no means a frequent occurrence. Very exceptionally, a male may display once more and drive the female for a short distance. Somewhat more frequently, the male may show a sudden and very temporary increase of aggressiveness after copulation expressed by aggression toward inanimate objects or other males if any are present. Such aggression toward other males may occasionally occur in any of the species under discussion. In blackbuck, they are very typical, and a suitor may even search for the nearest buck if none is already close to him. In Grant's gazelle, they are relatively frequent in a special situation. When a mixed herd has entered an open-plains territory and the owner has tolerated the non-territorial males before and during the mating ritual, he may start chasing them in a true outburst of aggression immediately after copulation.

BEHAVIORAL PECULIARITIES OF TERRITORIAL MALES AS COMPARED TO OTHER CLASSES

Differences in Flight Behavior

Provided that a territory is not too small and that the feeding conditions within it are sufficient, a "good" territorial male does not leave his territory readily. Consequently, he does not flee from predators, cars, or people as readily as do non-territorial conspecifics. Most instructive is a scene which frequently can be observed in the Serengeti plains where gazelle predators are numerous. Sometimes hundreds or thousands of grazing Thomson's gazelle are scattered over the plains. Usually, there are territorial males within such a concentration, but careful observation is needed to recognize them as long as the animals graze peacefully. However, the territorial individuals and their positions become strikingly evident the moment a predator suddenly appears close by but does not attack immediately. In such a case, all the non-territorial gazelle flee for some distance, but the territorial bucks remain in their territories and look toward the enemy. At this moment, the entire territorial mosaic becomes evident. It is as if a strong wind had blown all the leaves from a tree, and one suddenly sees the structure of the branches. If the predator starts hunting, the territorial males may also flee. In this case, there is a temporal difference in the flight behavior of the territorial individuals as compared to their non-territorial conspecifics. In other cases, spatial differences, i.e., differences in the flight distance (Hediger 1934), are more prominent.

Variations in the flight distances from predators, men, and cars according to age, sex, and social status of the gazelle can be recognized in all the species under discussion. Thomson's gazelle offer the best opportunity for quantitative and experimental investigation because there are many territorial and non-territorial individuals in a good tommy area, and the territories are relatively small and close together. Thus, one can easily test a statistically significant number of individuals from each class, and one can conduct the whole study within the same area excluding the objection that differences in the flight distances may simply be due to different situations of the animals in different areas. Of course, the flight distances also vary with the species of predators, their numbers, and other circumstantial factors (Walther 1969). However, differences in flight distances due to sex and social classes show up regardless of which predator or under which circumstances the gazelle flee. Thus, the principal points of intraspecific differences obtained by experiments with one flight releasing object can *mutatis mutandis* be applied to flight from other objects. In the experiments under discussion, the escape reactions of Thomson's gazelle from the observer's car were used for an investigation on intraspecific differences in flight distances. (For technical details see Walther 1969.) All the experiments were conducted in the Togoro plains within a period of two months at the same time of day (between 10:30 and 14:00), and all the data were taken singly and independent of each other. The tests included 100 territorial males, 100 non-territorial adult males in herds, 100 subadult males, 100 adult females, and 100 solitary wandering adult males. The latter were males who either had temporarily left their territories or, more frequently, had left a bachelor herd or become separated from it.

The largest flight distances occurred in the solitary wandering adult males (Figure 66). This result may be related to the insecurity of a gazelle which is neither in its territory nor together with conspecifics in a herd. The fight distances of adult females and of subadult males were significantly ($p<0.001$) larger than those of non-territorial adult males in herds, and the flight distances of the latter were significantly ($p<0.001$) larger than those of territorial males. Thus, the owners of the territories have the smallest flight distances of all the sex, age, and social classes investigated.

Whether these small flight distances make territorial males more vulnerable to predators, as has been suggested for territorial wildebeest bulls (Estes 1969), seems to be dubious in gazelles. At least in Thomson's gazelle, no evidence was found to support the view that the mortality rate by natural predation is greater in territorial males than in other classes (Walther 1969). One could even think of the opposite since in some predators, such as cheetah, which prey heavily on gazelle, the pursuit and the attack are apparently more easily released by a fleeing prey animal than by a standing one. Also, in the observation area in the Togoro plains, two hyena dens were right within territorial mosaics of Thomson's gazelle. When the hyenas left their dens in the evening, they always were surrounded by tommy bucks standing in their territories, but not the slightest hunting attempt was ever seen in this situation, although hyenas do hunt Thomson's gazelle. Moreover, in the three-year study on gazelle behavior in Serengeti, only one case was witnessed in which a territorial male was killed by predators.

Differences in Forms of Aggression

Since intraspecific aggression plays an important role in territorial behavior and also occurs in all sex, age, and social classes, it offers a particularly good possibility for demonstrating behavioral peculiarities of territorial individuals as compared with other classes. In principle, these differences are only quantitative; however, the numerical difference sometimes is so great that it approaches a qualitative difference. An example has already been mentioned in previous discussions: the grazing ritual of Thomson's gazelle is so frequent in territorial males, but so infrequent in non-territorial animals, that it can practically be used as an indication of territoriality. When one sees two adult tommy bucks in a pronounced grazing ritual and concludes that these two males are owners of territories, the possibility of an error is minimal. More differences in the agonistic behavior of territorial males as compared with non-territorial conspecifics will be pointed out in the following. Perhaps the most instructive method for such a comparison is to contrast the aggressive interactions of territorial males among each other to the intraclass encounters of non-territorial conspecifics.

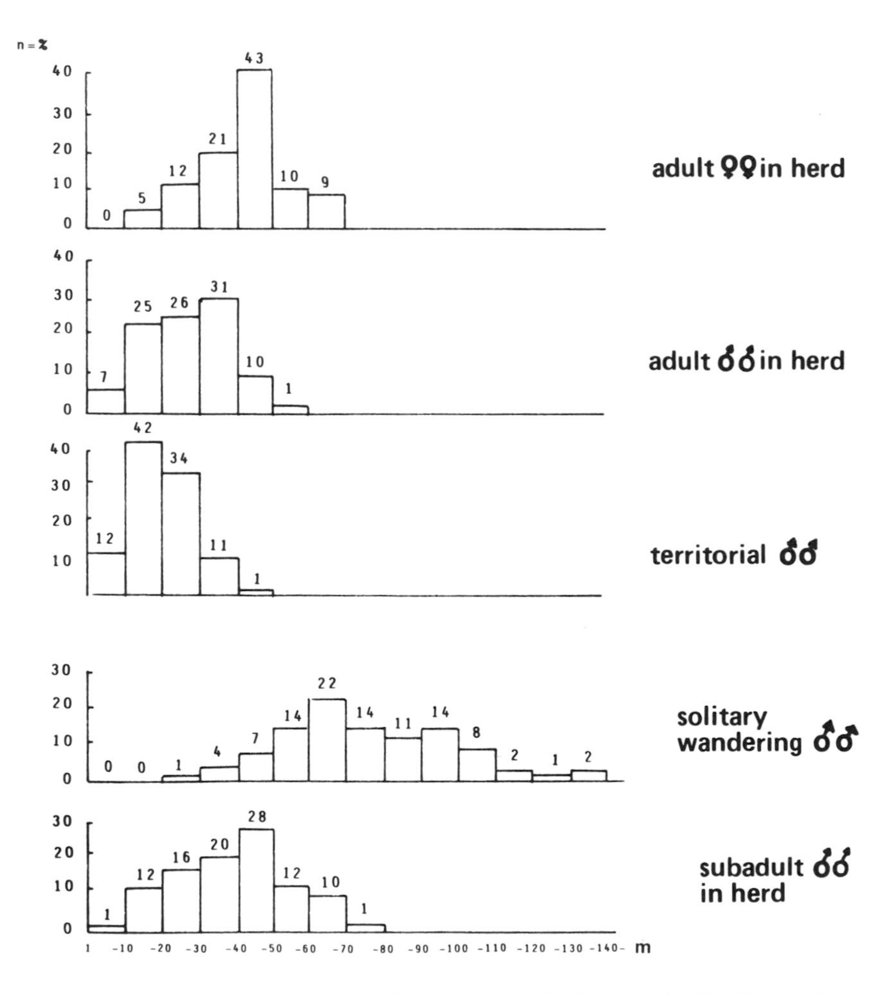

Figure 66: Flight distances of Thomson's gazelle from car in the Serengeti National Park (Togoro area in 1965). For details on data gathering and statistical evaluation see Walther 1969.

In mountain gazelle, Grau (1974) distinguished the following forms of aggression: direct approach, head-turned-away display, high presentation of horns, medial presentation of horns, fight, and chase (Figure 67). Despite the relatively small number of encounters observed among territorial bucks, a few differences from the agonistic encounters among non-territorial conspecifics are obvious. In the encounters among territorial bucks, the head-turned-away display occurred as the most frequent form of aggressive behavior, whereas it was infrequently used in the encounters among adult bachelors and lacking in the encounters among immature males and among females. In contrast, the medial presentation of horns was most frequent in all the non-territorial animals. Furthermore, the high presentation of horns almost equalled the medial presentation in the encounters of territorial peers. In adult and subadult bachelors, the high presentation was considerably less frequent than the medial presentation, and the high presentation was lacking in encounters among females.

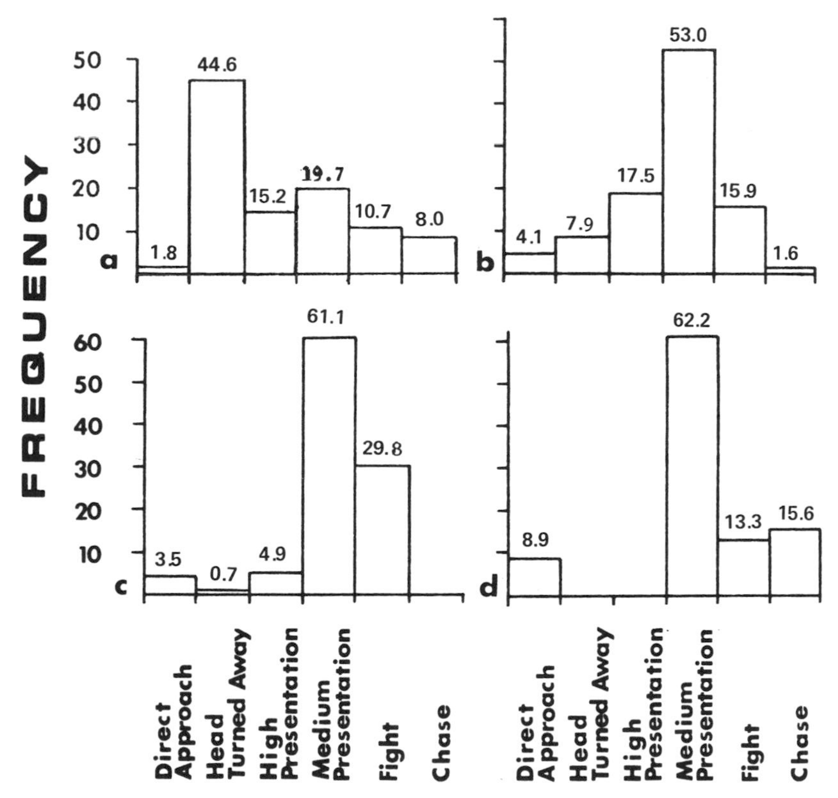

Figure 67: Occurrence frequency of forms of aggression in mountain gazelle opponents of different classes (in percentage). a = territorial male versus territorial male (112 occurrences of forms of aggression in 18 encounters); b = non-territorial adult male versus non-territorial adult male (434 occurrences of forms of aggression in 94 encounters); c = immature (subadult + adolescent) male versus immature male (283 occurrences of forms of aggression in 44 encounters); d = female versus female (45 occurrences of forms of aggression in 28 encounters). For more explanation see caption to Figure 55.

In Thomson's gazelle, Walther (1978a) distinguished the following forms of aggression: one-sided horn threats (including both high and medial presentation), reciprocal horn threats (including both high and medial presentation), air-cushion fights (only those cases in which no horn contact was made during the entire encounter), fights (with horn contact), body attacks, grazing rituals, pursuit marches (with medial or, usually, high presentation of horns on the part of the pursuer), and chases (Figure 68).

The encounters among territorial bucks were characterized by (a) reciprocal horn threats which usually were continued by fighting and thus, did not decide encounters, (b) fights which decided more than 50% of the encounters, and (c) grazing rituals which either followed after fights or occurred on their own (i.e., without any other form of aggression). Of the other forms of aggression, only object aggression occurred with a moderate frequency and was always followed by another, decisive form of aggression. All other forms of aggression were infrequent.

In non-territorial adult bucks, the situation was very different. Reciprocal threats were also followed by fights in most of the cases; however, they occurred in a considerably smaller portion of the encounters, and likewise, the proportion of fights was significantly ($p<0.001$) smaller than in territorial males. Grazing rituals were exceptional. Very much in contrast to the encounters among territorial males, the one-sided horn threat was the most frequent and effective form of aggression among adult bachelors, and was often followed by a pursuit march.

Similarly, one-sided horn threats were the most common form of aggression in encounters among adult females and decided many of them; however, here they were infrequently followed by pursuit marches. Reciprocal threats and fights were in the same small range as were all the rest of the forms of aggression. Grazing rituals were lacking.

The encounters among immature females largely followed the same pattern as those among adult females with an insignificant ($p>0.20$) increase in the proportion of fights. However, the proportion of pure air-cushion fights was significantly ($p<0.01$) greater than in the encounters among adult females.

The encounters among subadult males and among adolescent males show the relatively greatest similarities to those among territorial males in that reciprocal horn threats were frequent but were usually followed by other, more decisive forms of aggression, and that also the proportions of fights were great. (High and medial presentations of horns were lumped in this investigation. If they had been differentiated, a difference probably would have shown up in that high presentation is more frequent in territorial males, but medial presentation is more frequent in immature males.) Differences from the encounters among territorial males were the absence of grazing rituals and the relatively numerous pure air-cushion fights. (Air-cushion fighting is quite frequent in territorial males, however, not in pure form, i.e., it usually occurs before, after, or as an intermezzo within a fight with horn contact. These cases have been included in the fights with horn contact in this investigation.)

The encounters among juvenile animals (half-grown fawns, fawns, and neonates) differed from all the others in that threats were infrequent or even lacking, the numbers of fights and pure air-cushion fights which together made

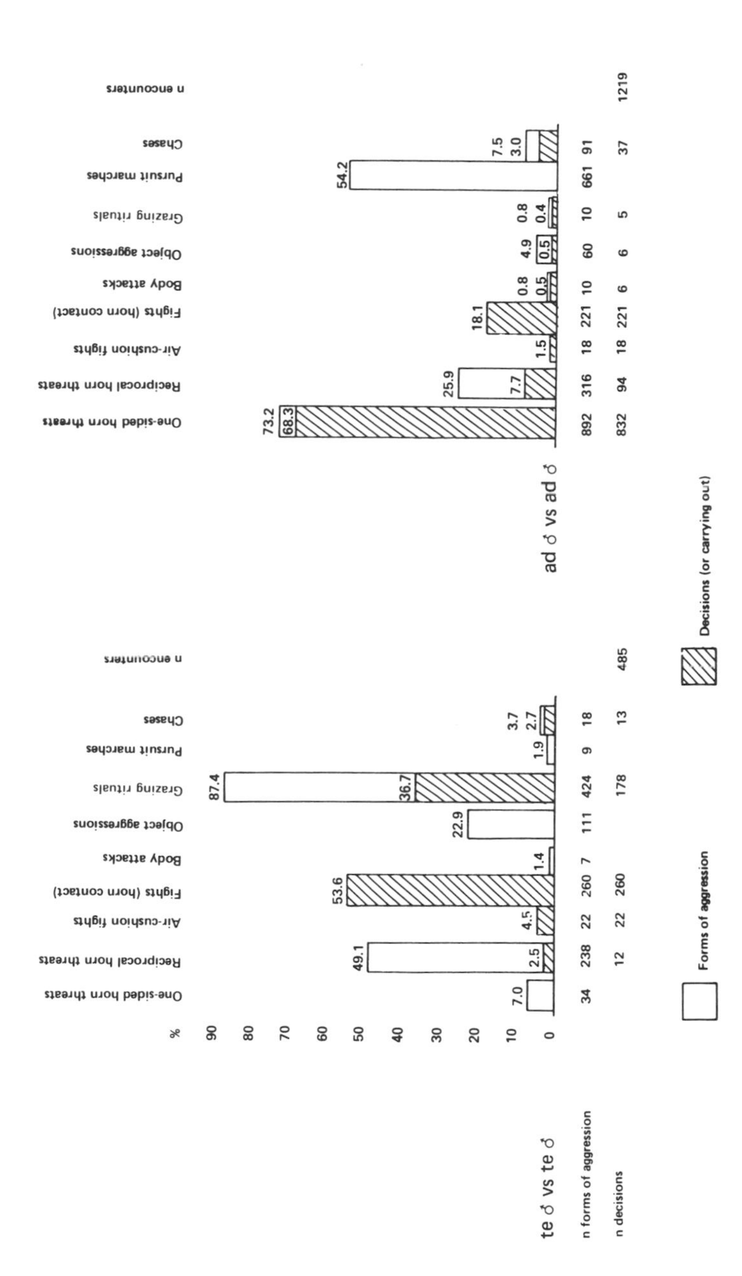

Figure 68: Forms of aggression and decisions (or carrying out) of agonistic encounters in different classes of Thomson's gazelle (in percentage). Abbreviations: im ♀ = immature (subadult + adolescent) female; hgf = half-grown fawn; faw = fawn. For other abbreviations and further explanation see caption to Figure 54.

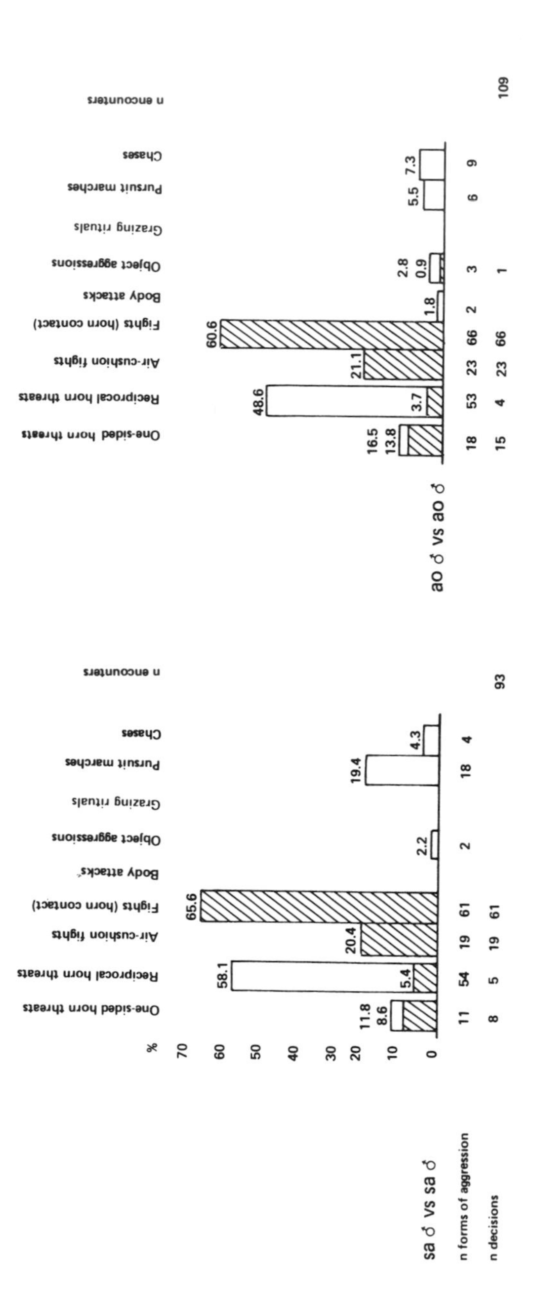

Figure 68: (continued)

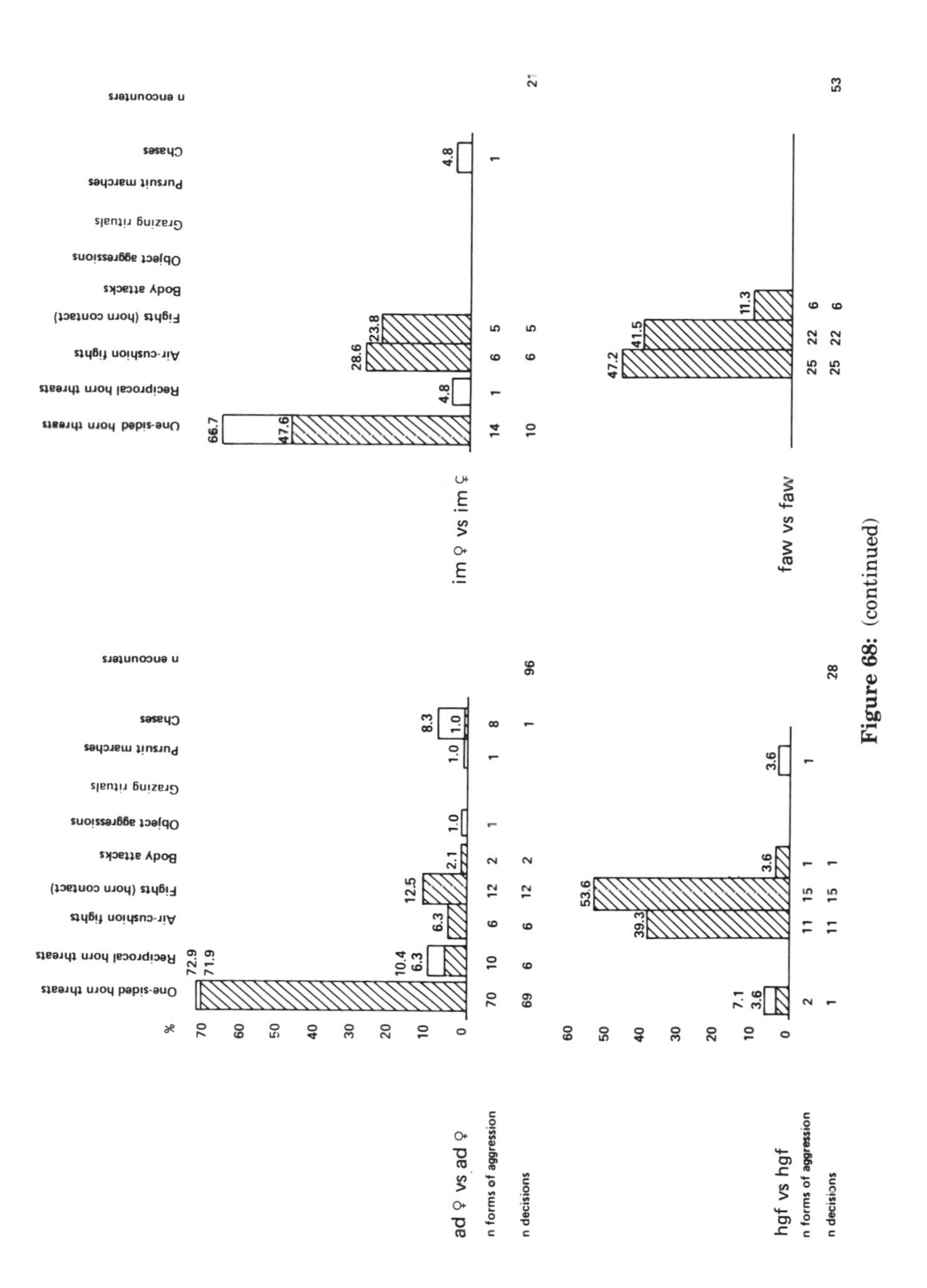

Figure 68: (continued)

up the bulk of these encounters, more or less equalled each other, and body attacks were comparatively more frequent than in the encounters of animals from any other class among each other.

A corresponding investigation of forms of aggression in Grant's gazelle brought similar results (Figure 69). As compared to Thomson's gazelle, quite interesting differences arose (a) from the greater elaboration of dominance displays in Grant's gazelle, and (b) from the lack of a grazing ritual comparable to that of Thomson's gazelle. In a sense, the tommy's grazing ritual substituted for the dominance displays of Grant's gazelle.

The following forms of aggression were distinguished in Grant's gazelle: one-sided walking display (walking straight toward the opponent without showing a special posture or gesture), one-sided head-flag, one-sided circling (with angled horns), one-sided threat (high or medial presentation), reciprocal walking display (i.e., both rivals approaching each other or, more frequently, marching parallel to each other), reciprocal head-flag (usually in reverse-parallel position), reciprocal circling, reciprocal horn threat, (pure) air-cushion fight, fight (with horn contact), body attack, object aggression, pursuit march, and chase (Figure 69).

As in tommy, one-sided threats and dominance displays were very rare in encounters among territorial Grant's bucks. Due to the greater elaboration of aggressive displays, reciprocal threat and dominance displays were even more frequent than in territorial tommy males. They often followed each other in a sequence, e.g., reciprocal approach and/or walking in a parallel march, reciprocal head-flagging in reverse-parallel position, reciprocal circling in reverse-parallel orientation with angled horns and finally high or, more frequently in Grant's gazelle, medial presentation of horns in frontal orientation toward each other. Again in contrast to Thomson's gazelle and due to the greater number and elaboration of aggressive displays—particularly the head-flagging and the circling display—a certain proportion of the encounters were settled by these displays which sometimes were followed by pursuit marches (with erect display on the part of the pursuer). In combination with the greater size of Grant's territories and frequently a broad strip of no-man's-land being between the territories, pursuit marches also occurred proportionally more often than in tommy. Reciprocal horn threats were usually followed by fights whose relative frequency was another characteristic of the encounters among territorial Grant's males. Object aggressions occurred proportionally more frequently than in tommy probably also due to their greater ritualization (weaving!), but they decided only a minor number of the encounters.

The encounters among non-territorial adult Grant's males were different from those among territorial bucks in that the one-sided threat and dominance displays played an incomparably greater role and decided many of the encounters. Correspondingly, reciprocal threat and dominance displays were proportionally less frequent in the encounters among non-territorial adult males; however, quite a proportion of the encounters among bachelors were settled by them, so that the proportion of fights was in about the same range as in tommy bachelors and was proportionally much smaller than among territorial males.

As previously mentioned, adult Grant's females are more aggressive than females of all the other Antilopinae species. Their aggressive interactions

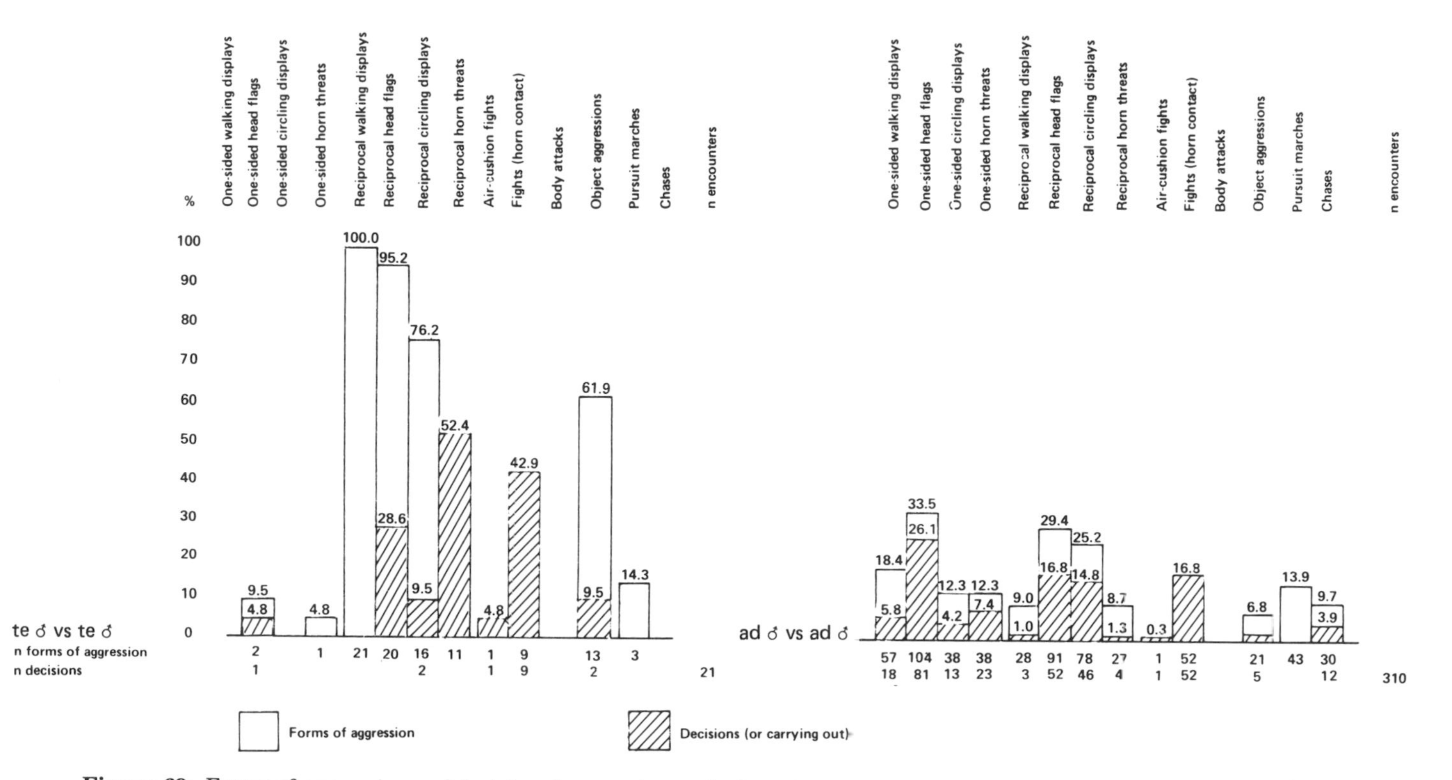

Figure 69: Forms of aggression and decisions (or carrying out) of agonistic encounters in different classes of Grant's gazelle. Abbreviations: juv = juvenile (neonate, fawn, and half-grown fawn). For other abbreviations and further explanation see caption to Figure 54.

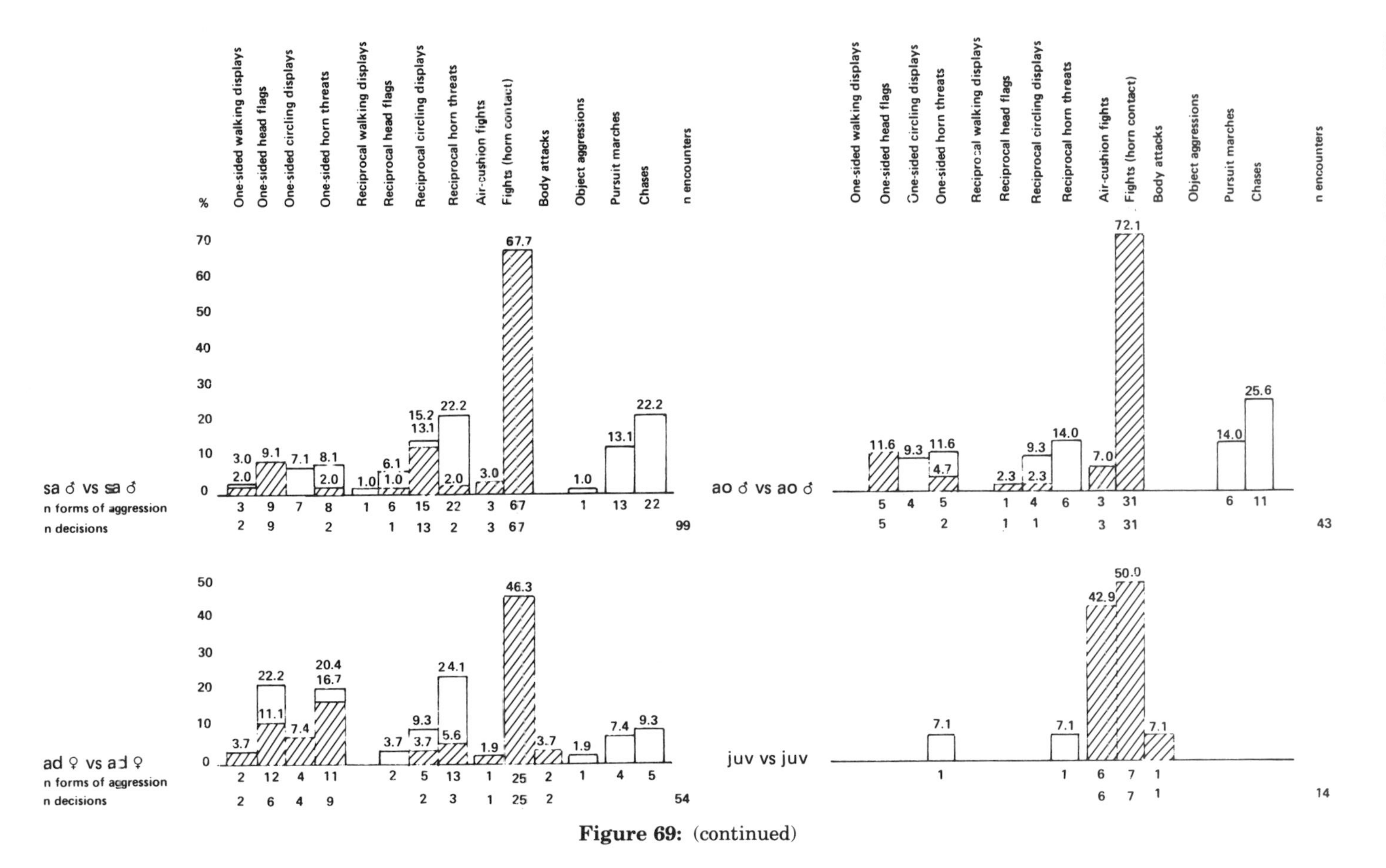

Figure 69: (continued)

essentially followed the same pattern as those of subadult and adolescent males and immature females. All of these differed from the agonistic encounters of territorial bucks in that one-sided displays were proportionally more frequent and reciprocal displays considerably less frequent. Furthermore, they differed from the encounters of both territorial and non-territorial adult males in that the occurrence of one-sided and reciprocal horn threats equalled or exceeded those of other (i.e., dominance!) displays. They differed from encounters among non-territorial adult males by the smaller proportions of encounters settled by one-sided and reciprocal threats and dominance displays, and the considerably greater proportions of fights which even exceeded the proportion of fights in the encounters among territorial bucks.

As in Thomson's gazelle, the encounters among juveniles were characterized by the rareness or absence of aggressive displays, the great proportion of fights and air-cushion fights, and the occasional occurrence of body attacks.

Additional Differences in the Behavior of Adult Males

Within the comparisons of territorial individuals to that of non-territorial conspecifics, the comparison to non-territorial adult males is of special importance because these behavioral differences are not due to the animal's belonging to different sex or age classes. They result purely from differences in the social status of otherwise equal individuals. Thus, the comparison of territorial males to adult males of different social status can teach us the most about the peculiarities of territoriality. Such quantitative behavioral differences linked to differences in social status are not restricted to the above examples of flight and aggression but may also occur in other behaviors.

In mountain gazelle, Grau (1974) put the frequencies of occurrences of certain behaviors in proportion to the time spent observing the individual animals (individual-hours). These proportions were compared in territorial and non-territorial males (Table 8) for several forms of aggression (chasing, broadside displays, fights) driving females, urination-defacation sequences with (S-U-D) or without (U-D) scraping the ground preceding it, and weaving (object aggression). Driving of females did not occur in the bachelors observed. Particularly impressive were the greater frequencies of urination-defecation sequences and weaving in the territorial bucks, a result which was expected since these behaviors can serve to mark territories. Also, all the other activities were more frequent in the territorial males except fighting because fights predominantly take place among peers, and the territories of mountain gazelle are very large and often not directly bordered by neighboring territories.

Both in Texas and in India the round-the-clock activity during daylight hours—7:00 through 19:00 for Texas and for India—of a territorial blackbuck male was compared to that of a bachelor association observed at the same study site. The four basic activity categories distinguished were: lying, standing, grazing, and active without grazing. As illustrated in Figure 70, the activities of territorial bucks and non-territorial males at the same site followed principally the same pattern of changes between quiet periods and periods during which the animals were active. Also, the total times which the territorial bucks spent in standing and in grazing at each site were about the same as for the corresponding bachelors: 7.2% standing and 37.6% grazing (territorial male) versus 7.0% standing and 36.0% grazing (bachelors) for Texas, 11.9% standing and 47.0% grazing (territorial male) versus 15.4% standing and 51.8% grazing (bachelors) for India (Point Calimere). Time spent grazing was the greatest percentage of any category for all blackbuck—including the females in female groups—at both sites. Lying time was greater for the bachelors in Texas (23.7% in territorial male versus 38.9% in bachelors) but similarly low for all males at the Indian site (11.2% in territorial male versus 10.2% in bachelors). Most conspicuously, the territorial males, with their herding and courting of females and displaying to bachelors, were nearly two to three times more active than the non-territorial males (31.5% in territorial male versus 18.1% in bachelors in Texas, 21.9% in territorial males versus 8.5% in bachelors in India).

Table 8: Comparison of Behavior Patterns Used by Territorial Mountain Gazelle Males and Bachelors During Daily Activity Watches

	. . . . . Territorial Males		. . . . Bachelor Males.	
	Total No. of Observations	No. of Observations per Individual-Hour	Total No. of Observations	No. of Observations per Individual-Hour
No. times driving females	68	0.448	0	0.000
No. S-U-D or U-D	102	0.672	12	0.029
No. times chasing males	29	0.191	12	0.029
No. broadside displays with head-turned-away	31	0.204	1	0.002
No. times weaving	87	0.574	29	0.071
No. of fights	6	0.040	54	0.132
Total no. of males observed	14		42	
No. of individual-hour observations	151.7		410.46	

Note:
The individual-hour is one individual observed for one hour.
S-U-D = Scraping-Urination-Defecation Sequence.
U-D = Urination-Defecation Sequence.
Weaving = object aggression.

In Thomson's gazelle, an adult buck can be territorial or a member of a bachelor group or a member of a mixed (migratory) herd. The frequencies of certain behaviors (marking with preorbital glands, urination-defecation sequence, object aggression, horn threat, fight, grazing ritual, chasing a male, chasing a female, neck-stretch, nose-up including nose-lift and *Laufschlag* in combination with lifting the nose, *Flehmen, Laufschlag* in normal posture, mounting, and copulation) were recorded in 10 males of each class. Each male was observed for 12½ hours, from 6:30 through 19:00 (Figure 71).

When one compares the territorial bucks (Figure 71a) to the adult males in bachelor herds (Figure 71b), one realizes immediately that all the behavior patterns under discussion occurred more frequently in the territorial males,

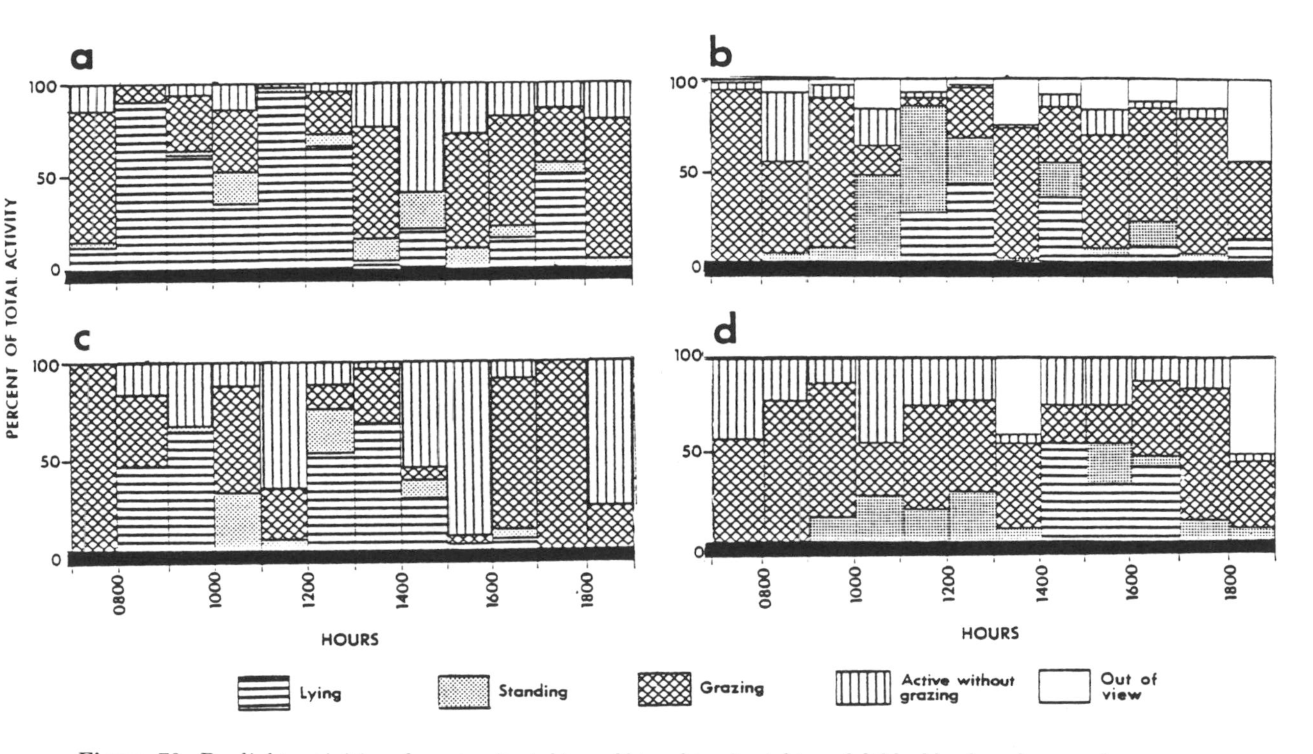

Figure 70: Daylight activities of non-territorial (a and b) and territorial (c and d) blackbuck males. a and c compare a bachelor group and a territorial male in the same Texas pasture observed within the same week. b and d compare a bachelor group and a territorial male in the same area and on the same day in India.

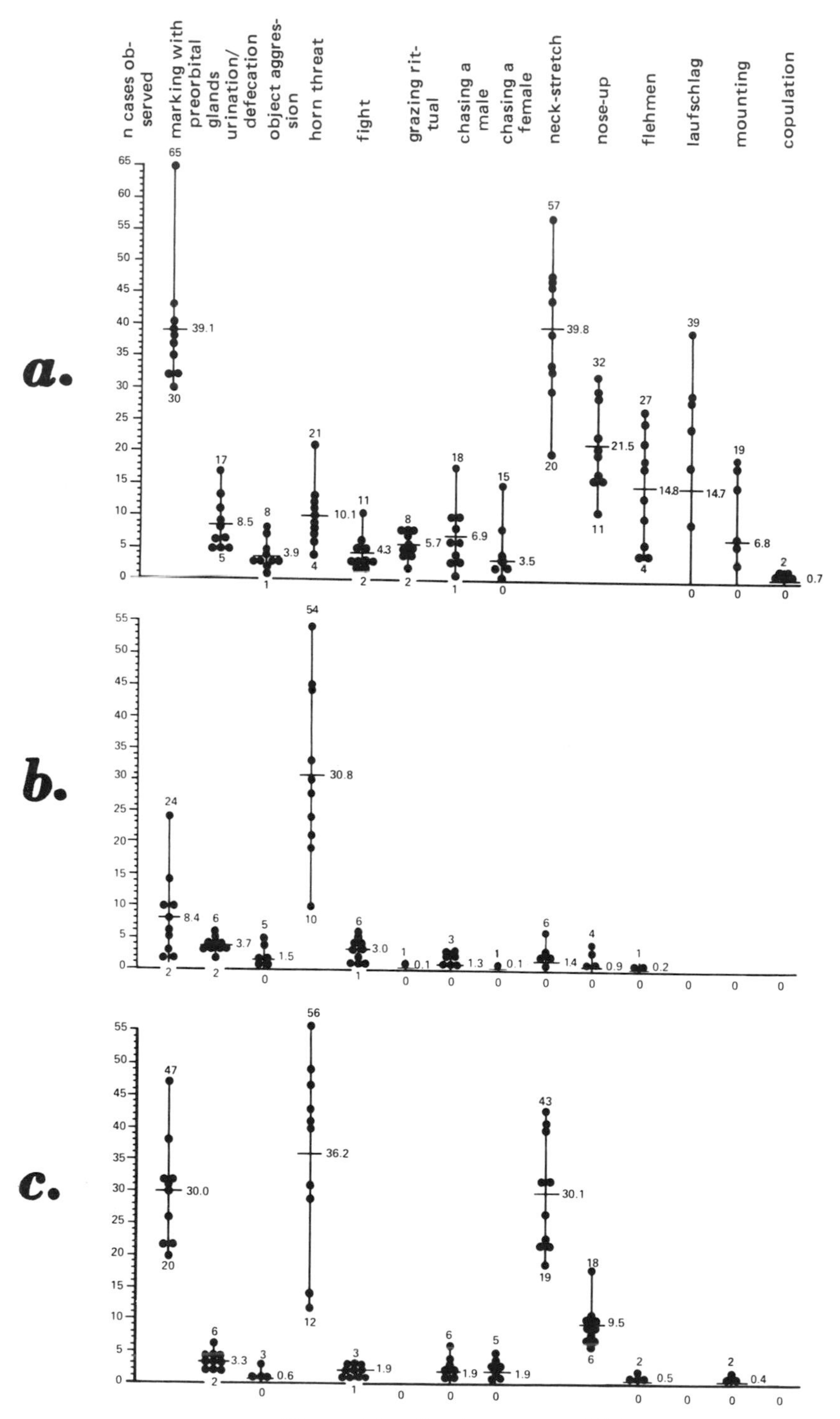

Figure 71: (Caption on page 167).

Figure 71: (a) Frequencies of certain behavior patterns in 10 territorial males of Thomson's gazelle (during 12½ hours). The black dots indicate the frequencies in which the activities under discussion were observed in each of the 10 males. No dot was made if a male did not show the corresponding behavior pattern. The vertical lines indicate the quantitative range of each behavior pattern. The numbers on top and bottom are the maximum and the minimum numbers observed. The horizontal lines with numbers to the right indicate the means. For more information on data collecting and statistical evaluation see Walther 1978b. (b) Frequencies of certain behavior patterns in 10 adult males of Thomson's gazelle in (stationary) bachelor herds (during 12½ hours). (c) Frequencies of certain behavior patterns in 10 adult males of Thomson's gazelle in (mixed) migratory herds (during 12½ hours).

with one exception: horn threats, i.e., mainly high and medial presentation of horns, were more frequent in the bachelors. Behavior patterns indicative of the advanced phases of a mating ritual (*Laufschlag,* mounting and copulation) were missing completely in the observed bachelors. With the exception of fights ($p>0.10$), all the differences were found to be significant ($p<0.025$ to <0.005, see Walther 1978b). The greater frequency of horn threats in adult bucks in all-male groups was due to the frequent encounters for coordination of group activities which, of course, were lacking in territorial individuals.

When one compares the behaviors under discussion in territorial bucks (Figure 71a) to those of non-territorial adult bucks in mixed herds (Figure 71c), essentially the same picture emerges. All the behaviors, except threats, were more frequent in territorial males, and again, all the differences (including fights) were found to be significant (Walther 1978b), except for chasing a female ($p>0.10$). However, the differences were often not as great as those with males in the bachelor herds (Figure 71b). Grazing ritual, *Laufschlag,* and copulation were lacking in the adult bucks in mixed herds, and quite a number of the behavior patterns under discussion were in about the same range as in the bucks in the bachelor herds. However, the marking activity was greater in the migratory males, and above all, some behaviors typical of the association with females (such as neck-stretch, nose-up, and chasing a female) occurred more often in the bucks in mixed herds. Even a few spontaneous mounts (i.e., mounting without previous courtship displays) were observed. This greater frequency of interactions with females is not only because females are permanently present in the mixed herds, whereas bachelor groups may only occasionally meet females. Bucks in the mixed herds perform the neck-stretch and the nose-up for the same purpose as the (frequent!) horn threats toward males; to make the females switch from resting to moving, to speed them up when they slow down during a move, etc. In short, threats toward other males and "courtship" displays toward females are used to coordinate the group activities by non-territorial adult tommy males in mixed herds.

SPECIAL RELATIONS AMONG TERRITORIAL NEIGHBORS

The presence or absence of neighbors probably has a certain influence on territorial behavior in all the Antilopinae species. However, this influence ranges from mildly important to extremely important. Naturally, the importance becomes greater as the territories become smaller and denser. Among the Antilopinae species investigated to date, gerenuk and Thomson's gazelle seem to represent the two extremes. In a three-year study, Leuthold (1978a) apparently did not see any interactions between territorial gerenuk neighbors with their large territories in comparatively densely wooded areas. One may assume that such encounters take place occasionally, but they definitely are rare in this species. Thus, the relationships between territorial males do not play a great role in gerenuk, except of course, that when a male establishes a territory he has to find a place not occupied by another male.

Thomson's gazelle have relatively small territories in the open plains which are surrounded by neighboring territories. The interactions with neighbors are so important to the territorial status of these males that, with little exaggeration, one may say that a territorial buck without neighbors is not a territorial buck in this species. The situation in blackbuck is close to that in tommy. Mountain gazelle, springbok, and Grant's gazelle seem to be more or less in between the extremes characterized by gerenuk and tommy.

In species where neighbors play a prominent role in territoriality, it is justifiable to speak of a "dear enemy" relationship (Fisher 1954) among the territorial individuals. One must keep in mind that all Antilopinae species are more or less gregarious, some of them even very gregarious. On the other hand, when becoming territorial, a male leaves the herds and more or less isolates himself from the others. Even when his territory is regularly frequented by females, they usually stay with him only a few hours a day. Trespassing bachelors are chased away, and they may eventually learn to avoid the occupied area when they stay in its vicinity long enough. Thus, even under the very best conditions, a territorial buck is alone for several hours every day. If the area is not readily frequented by females, he may be alone in his territory for days and weeks, and such cases are by no means unusual under natural conditions. Then, the encounters with territorial neighbors are the only social interactions for such a male. Thus, the presence of neighbors may be an essential factor in satisfying the social needs of such a basically gregarious animal in spite of its territorial isolation.

There are indications that the territorial neighbors may know each other individually. Apparently, this individual "acquaintance" can either result in a diminution or an increase of the aggressiveness of the bucks toward each other. How and why it works the one way or the other is largely unknown. For example, when a territorial tommy buck has four or five immediate neighbors, he can be strikingly less aggressive toward one or two of them than toward the others. Of course, he has boundary encounters with all of them. But the encounters with "his friends" are relatively infrequent, often without physical contact, i.e., settled by mere grazing rituals and/or air-cushion fights, and, if they lead to fighting, the fights are often over after one or only a few clashes. He may even sometimes tolerate it when such a "friend" crosses the boundary —at least as long as the latter does not deeply intrude upon his territory. With other neighbors, he may have many encounters usually resulting in fights of considerable length and vehemence, and he may charge them as soon as they show up near the boundary.

Even more frequent are cases in which a stronger aggressiveness indicates that the neighbors know each other. When territoriality is declining (see p. 184), a territorial tommy buck may tolerate non-territorial males inside his territory and sometimes even accompany them some distance as they travel ahead. In this situation, a territorial buck may leave his territory together with bachelor groups of 30 or more adult males and may enter his neighbor's territory. Provided that the latter is also in a declining phase of territoriality, he may tolerate all the bachelors, but he promptly charges his neighbor.

Likewise, a territorial Grant's buck, in an open-plains territory (p. 176), may sometimes leave his territory together with a mixed herd which has been visiting him for some time. In contrast to Thomson's gazelle, apparently this situation does not have anything to do with a decline of territoriality in Grant's gazelle. Outside his territory, such a male does not behave differently from the other non-territorial males in the mixed herd. Nevertheless, when the herd arrives in the territory of the next territorial buck, the latter usually tolerates and only mildly dominates the non-territorial males but often may intensively display toward his neighbor and may drive him out of his territory. Sometimes he may even fight him.

THE DAY OF A TERRITORIAL BUCK

In the interest of a better presentation and discussion, the previous chapters, in a sense, have cut the activities of a territorial male into pieces. Therefore, it is perhaps not out of place to complete the picture by putting the pieces together in a more holistic approach. The most instructive way may be to describe a typical day of a male during the peak of his territorial status. Since such descriptions easily become lengthy, one example may suffice. We take the example of a territorial tommy buck named "Roman" in the Togoro plains describing a "typical day" as it emerged from numerous observations of this individual.

The last hours before daybreak (about 6:00) "Roman" has spent alone resting in his territory. With the first light, he rises and takes a "marking walk." He walks to a dung pile, scrapes, urinates and defecates there, and he renews secretion marks on grass stems along the boundary; of course not all of them, but quite a number. In pauses during the "marking walk," he may gore grass and ground with his horns. Then he begins to graze, frequently close to the boundary; a situation which may lead to the first agonistic encounter (boundary ratification) with a neighbor. Provided that no females have come into his territory yet, he stops grazing between 8:00 and 9:00 o'clock, and beds down in the open with his back toward sun and wind. If females have arrived, he remains on his feet. (This happened infrequently because the females usually approached the periphery of the territorial mosaic in this area in the morning, and "Roman's" territory was located in the center of the territorial mosaic.) The morning rest during which the animals obviously warm up after the cool night is very obligatory in gazelles, and it usually lasts for one to two hours. The females, which may have arrived in the peripheral territories, also rest during this time. They arise between 10:00 and 11:00 and move ahead while grazing. They now approach "Roman's" territory. He waits for them at the boundary. As soon as the first female has passed the boundary, "Roman" positions himself behind her and, displaying neck-stretches, nose-lifts and nose-ups, some combined with drumrolls, he drives the female toward the center of his territory. He may have to repeat this procedure with more females, but finally the rest of them follow the first female without being herded by the male. In this way, easily 20 to 30 but sometimes 100 or more females may gather in "Roman's" territory.

He now approaches one of the females with neck-stretch and nose-up and drives her until she urinates. He sniffs at the urine on the ground and performs *Flehmen.* If she is not in estrus, he may test more females, one after the other. If "Roman" finds a female in the "right" state, she walks ahead while he performs *Flehmen.* He approaches her again alternately displaying neck-stretch and nose-up. The female walks in a zig-zag course, sometimes she even turns 180°, but "Roman" follows her persistently and begins to treat her with foreleg-kicks.

Suddenly, "Roman" stops and watches. A non-territorial buck has entered his territory and is approaching the females. "Roman" turns away from his favorite, "politics before love," and, displaying high presentation of horns, he walks straight toward the bachelor. The latter sees him come, lowers his head to the right and the left, lifts his feet alternately as if the ground had become hot, grooms his inguinal region, vigorously scratches his neck, shakes his flank, and finally flees at a gallop.

"Roman" returns to his chosen female and starts the mating ritual from the beginning. Eventually, the ritual merges into a mating march in which he walks close behind her without any display. The pair moves in a zig-zag course at the periphery of his territory. On the other side of the boundary, neighbor "Short-Tail" grazes alone in his territory.

Finally, the female stretches her tail horizontally, indicating that she is now ready for copulation. "Roman" mounts her while walking. After several mounts, he copulates. For a moment, both mates stand quietly, then walk away from each other in different directions.

Meanwhile, the other females have crossed "Roman's" territory and are now leaving it. "Roman" hurries toward them and tries to herd back at least some of them but he is not very successful. Eventually, only one female is still with him, and she too tries hard to leave the territory. "Roman" herds her back toward the center, but she suddenly turns and runs toward the boundary at full gallop. "Roman" pursues her as fast as he can. He passes her and blocks her path in broadside position. The female stops, turns, walks back a few steps, and again tries to break through at a gallop. "Roman" can stop her a few more times, but finally she crosses the boundary. "Roman" blindly follows her in a wild chase and runs far into the neighboring territory. The neighbor charges him and they clash together, both from a full gallop. After several violent clashes, "Roman" returns to his territory.

Meanwhile, more females have come from the other side and have just arrived in neighbor "Short-Tail's" territory. "Short-Tail" courts one of them intensively but this female is not very cooperative. She does not walk in front of the male. Instead, she trots and gallops and frequently turns sharply. Eventually the buck "loses patience" and chases her at a full gallop. "Roman" stands precisely on the boundary and watches the chase. When the female enters his territory, he takes over and chases her across his own territory. This time he respects the boundary and stops from full gallop in a cloud of dust. The next territorial male takes over and chases the female ahead. She is now breathing with open mouth and is obviously exhausted. When she slows down to a walk, the buck starts courting and driving her. After a short march, she stretches her tail horizontally. The buck copulates after several mounts.

At high noon, "Roman" is alone in his territory. There is no tree in it, but there is a tree only a few yards from the boundary in a neighboring territory. "Roman" walks to this tree and stands in its shade during the hottest hour of the day. The neighbor tolerates his presence in this case. He has another tree in the center of his territory where he stands and dozes.

After about an hour, "Roman" returns to his territory. The afternoon passes with grazing and marking. About 17:00, "Roman" beds down for half an hour. Then he grazes again. The females now return from the direction in which they had disappeared at noon. Upon their return, the herding, testing, and courting activities of the territorial males start again. However, the females now come from an area not occupied by territorial males where they have intermingled with a bachelor herd, and quite a number of non-territorial males are still with them. With loud vocalizations, but without any other display, "Roman" rushes toward the bachelor next to him and chases him out of his territory. Then he charges the next one, and so on. His neighbors are also chasing the non-territorial males, and soon the female herd is "cleaned out."

As the sun sinks, "Roman" has another boundary ratification encounter with neighbor "Short-Tail." He walks straight to the boundary and stands there with highly presented horns until "Short-Tail" becomes aware of it, accepts the challenge, and approaches. After a reciprocal threat display, they clash together in a fight followed by a long grazing ritual. Such late-evening encounters are obligatory among territorial neighbors, and they also take place when no females are present.

The herding of females, chasing of bachelors, and encouters among the territorial neighbors are continued during the short tropical dusk. After nightfall (about 19:00), the gazelles are still active for about half an hour. Then they bed down for rest. Incidentally, they spend the greatest part of all resting periods in dozing and ruminating, but they sleep for only very short periods at long intervals. Five to 10 minutes of uninterrupted sleep is a long sleeping period for an adult gazelle. All the short sleeping periods in a 24-hour day added together, may total one to two hours; whereas these animals may easily spend eight to 10 hours lying.

About 22:00, the rest is usually interrupted by a short activity period with moving and grazing. This activity bout probably is induced by true or false alarms. Many predators are active at this time. After about half an hour of activity, the gazelles bed down again.

Another well-pronounced nocturnal activity period starts shortly before midnight. On dark nights, the animals move and graze for about an hour. However, this "midnight activity" can be continued for up to four hours on bright moonlight nights, and all the activities typical of territorial males (marking, herding, chasing, fighting, etc.) occur. Toward the end of this activity period, the females have usually left the area of the territorial mosaic. "Roman" is alone in his territory. He beds down and rests from about 4:00 until daybreak.

8

Variations in Territorial Behavior

As is evident from the previous chapters, Antilopinae territoriality is a rather complex matter composed of marking behavior, aggression, sexual behavior, etc. As a rule of thumb, one may say that behavior is the more variable as it becomes complex. This is also true for territoriality. Studies on territorial behavior in bovid species such as Uganda kob (Buechner 1961, Leuthold 1966) and wildebeest (Estes 1966) have revealed remarkable differences in the territorial behavior within the same species. Such intraspecific differences in different areas may be related to differences in population density, vegetation and other environmental conditions. It is likely that corresponding variations also exist in the territoriality of Antilopinae; however, not many of them have been thoroughly investigated. Nevertheless, some variations of this kind were observed in some of the species under discussion.

In Thomson's gazelle, variations in territorial behavior appear to be relatively insignificant under natural conditions. There are some differences but most of them either result from different phases of territoriality (see p. 178), or are such immediate consequences of special, momentary and/or local conditions that one may hesitate to speak of true variations in territorial behavior. Obviously, the relatively greatest variations occur in the length of the territorial periods (p. 60) in this species. As previously mentioned, territoriality is not restricted to a definite season in Thomson's gazelle. Consequently, there are always males who try to establish territories at any time and where they just happen to be. Since this species performs regular and remarkable migrations in Serengeti, some males try to become territorial during migration periods. If such a territory is on a major migration route and the migration is continuing, hundreds and thousands of gazelles may pass through it each day. Then, the territorial buck consumes his energies in fruitless attempts to herd females in his territory and to prevent non-territorial males from entering it. Eventually, such a buck is completely exhausted, and, usually after a few hours, he abandons his territory.

Or, a tommy buck may try to establish a territory when the herds are passing through an area which is not very suitable for gazelles in general or for a territorial male in particular, e.g., in a relatively dense forest. Provided that his territory is not "flooded" by the migratory herds and that he is not "carried away" by them, as described above, it is unlikely that other gazelles, particularly females, will stay there because they prefer open short-grass plains. Thus, soon he will be the only gazelle in the whole vicinity, a situation which clearly is unfavorable for the territorial status of a tommy male. Moreover, insects are much more abundant in tropical forests than in open areas, and gazelles are rather sensitive to them. When intensely bothered by insects, a tommy may even flee at a full gallop. This sensitiveness toward insects may be another reason why a tommy buck does not remain territorial in a forest area very long.

During a non-migratory period when a male tries to establish a territory near a river or another source of water during the dry season, he also will not be successful for long. In this case, large herds of males and females pass through his territory on their way to and from the water, and he soon exhausts his energies like a territorial buck on a major migration route. These and similar situations may account for certain "variations" in the lengths of territorial periods in this species.

As previously described, a tommy buck may annex parts of a neighboring territory after the neighbor has left and enlarge his territory in this way. Or, he may try to establish a very small "mini-territory" in a "no-man's-land" between neighboring territories. Besides such special and usually very temporary events, however, no striking variations in size of territories in different areas have been recorded in this species.

Another "variation" may arise from the presence or absence of trees inside a tommy territory or its immediate vicinity. As previously described, a buck may incorporate such landmarks in the structure (e.g., boundary line and/or center) of his territory, and he may spend the hottest hours of the day in the shade of such a tree. Of course, he cannot do so when no tree is available.

When large female herds slowly pass through a mosaic field of territorial tommy bucks, it may exceptionally (five observations within three years) happen that a territorial buck is "carried away" and moves together with them, leaving his territory and entering the neighbor's territory. However, he does not find any resistance there because his neighbor has also been "carried away," and likewise the neighbor's neighbor, etc. Thus, these males move with the females but keep their "territorial distances" among each other within the moving herd. Such an event regularly ends as soon as the "leading" territorial buck enters a territory whose owner does not participate in the move. Then, this owner charges the territorial male moving with the herd and, with or without fight, chases him back. In fleeing, this male runs into the moving territorial buck behind him, charges him and chases him back, etc. It is like a chain reaction, and, eventually, each of these moving territorial males is back in his own territory again.

A last "variation" of territorial behavior in tommy was observed only once, and it was clearly linked with the decline of territoriality (see p. 184). During the dry season, an area was left by most of the gazelles, but a few territorial males remained. For two weeks, these males regularly left their territories in the morning hours and gathered on an unoccupied area in the vicinity forming an all-male group. They remained there and behaved like bachelors until the late afternoon. Suddenly, as if a signal had been given they dispersed at a gallop, each of them returning to his own territory where he stood until the next morning.

As can be seen from these descriptions, "variations" of territorial behavior are more or less individual matters, often of almost anecdotal character in Thomson's gazelle. However, they may offer some starting points for a better understanding of the more pronounced variations found in certain other species.

In mountain gazelle, there also were only a few and not very striking variations of territorial behavior in different areas. The territories in the rough terrain of the Ramat Yissakhar area, with a relatively dense population, appeared to be smaller on the average than the territories in the flat terrain with a more scattered population in the northern Negev. Certainly, many of the dung piles were larger on Ramat Yissakhar than those in the Negev. The reason for this difference is unknown. Possibly the smaller average size of dung piles in the northern Negev simply was due to the frequent destruction of older dung piles by agricultural practices in this area. Before the mountain gazelle in Israel were protected by law, they frequently were observed in pairs or in "families" (male, female, young). It is very likely that this was an effect of the extremely small total population of gazelles at that time, i.e., a (presumably) territorial buck often could not find more than one female, possibly with her fawn, in the vicinity (Mendelssohn 1974).

In Indian blackbuck, variations of territory size, of length of stay, and with season have been described in previous chapters. Environmental complexity in particular is linked to territory size with the extreme in small territory size–the tightly clustered territorial mosaic at Velavadar–conspicuous in the flattest, most open country used by very large herds. Another way that environment influences native populations is by the degree to which monsoon conditions affect the territorial males. Point Calimere males are forced out of their territories by coastal flooding in June to August during South India's southwest monsoon, but they re-establish themselves in the same area after waters recede (Hussainy pers. comm.). In contrast, the inland males near Mudmal in central India keep their territories but neglect their dung piles from late June after the rains start until late November when the ground begins to dry (Prasad 1981). Due to the various ways in which introduced blackbuck are kept on Texas ranches, several further and comparatively well-pronounced variations of territoriality occurred in this study, most of them obviously related to differences in the space available to the animals.

Most significant are the effects of limited space (Mungall 1979). Regardless of animal density, small, zoo-type pens can accommodate only one territorial male. The pen boundaries serve as the territory boundaries. Since individuals cannot enter or leave such a pen-territory, the animals live together as one herd of "master buck," subordinate bucks, females and young. Without room for separation, there can be no bachelor association and no cycle of entry and exit of female groups. A human approaching the fence may be displayed to as if he was a territorial rival. Neonate fawns may be attacked by the master buck so fiercely that they die of their injuries (Benz 1973). For the most part, however, the master buck discharges his aggressive impulses toward the older of whatever blackbuck males are present. Lacking a male scapegoat, the master buck will direct his aggression toward his females. An unreceptive female may be courted relentlessly with liberal intervals of chasing. Eventually, the female may answer his mount intentions by boxing his shoulder, but he persists. In areas large enough for the animals of different sexes and social categories to separate, territorial bucks still chase and thrash during courtship, but unreceptive females escape and females hardly ever direct an aggressive gesture toward an adult male.

In medium-size pastures (roughly 20 to 200 ha), the full range of territorial behavior may be shown unless there is only one open area suitable for the establishment of territories. If there is only one such opening and if it has little to interrupt a buck's line of sight, a particularly aggressive male may monopolize nearly the whole opening as his territory. The females can move out of this territory, but they generally do not. Occasionally, they make forays to flee a disturbance or to visit water. Nevertheless, excessive chasing and impulsive courtship by the bachelors who inhabit the fringes of the territory guarantee that the females return promptly even if the territorial buck does not drive them back. The one territorial male does all the breeding. Such situations have encouraged observers to conclude that blackbuck keep harems when, in actuality, this variation of territorial behavior is only a special case resulting from space limitations.

When there is only one territorial buck in a pasture with a bachelor group, aggressive interactions with the bachelors substitute for the border encounters a territorial buck would normally have with his territorial neighbors. Just as in the small pen the master buck singles out the other males for aggressive approaches. An owner observed in a medium-size pasture with one territory habitually left his territory in the afternoon to go find the bachelors if they had not yet offered themselves as targets by trespassing that day. The confrontations satisfied the territorial buck's urge for aggressive interactions, and the buck's relations with the females remained normal.

Although it is typically the encounters between neighbors that keep territorial boundaries clearly defined, the driving out of trespassing bachelors also has a reinforcing effect. A buck who must expel bachelors periodically but has no territorial neighbors, observes his boundary as precisely as the same buck with territorial neighbors. Lacking either, a buck not limited noticeably by fences or natural physiographic barriers, drifts. His core area can be about the size that can be held by an aggressive buck with bachelors near but with no territorial neighbors (19.6 ha as opposed to 18.7 ha, respectively, in the case of two bucks observed). The buck without male company wanders off with females as they leave his core area. Eventually, he returns alone. As with the territorial buck trespassing on territories from which the owners are absent, the buck manifests no behavior changes as he leaves. Without challenges from other males, the buck indicates to the observer no line or strip beyond which he can be made to retreat.

When faced with only a few bachelors, a territorial male can keep the bachelors away from supplemental feed such as pellets and hay in his territory. Faced with large crowds of animals that include nearly half a dozen or more adult males, the owner begins by frantically displaying to first one and then another, but he cannot drive away the intruders. After the initial flurry of activity, he often settles down and gives more attention to the females and the feeder. However, most of the feed is gone by then. Since territorial backbuck normally often leave their territories to feed during part of the day, a territorial buck in a pasture where the animals are maintained on supplemental feed will finally get his chance at the protein pellets by leaving his territory and joining the crowd of animals at a trough elsewhere.

In the wild, the greatest variation, apparently due to two different types of environment and population density, was found in the territorial behavior of Grant's gazelle. In this species, bucks may establish their large territories on the *mbuga* within forest areas. These *mbuga,* as well as the surrounding forests, may be crossed by numerous Grant's gazelle during migrations but only some of them stay for a longer time. When the migration is over, almost no Grant's gazelle remain in the forests. Thus, during a non-migratory period, a relatively small population of Grant's gazelle on such a *mbuga* is isolated for miles from other Grant's gazelle by the forests. In this situation, the territorial bucks are as intolerant of other males as are the males of Thomson's gazelle at the peaks of their territorial periods. In contrast to the tommy males, however, a male Grant's gazelle keeps a relativey small (usually 10 to 20 members) but also relatively stable group of females within his territory for weeks and months. Thus, territorial intolerance is well-pronounced, herding the females (including "silent herding"—see p. 140) is frequent and successful, and, on the whole, territorial behavior and harem behavior are combined in this case.

In the same species, the situation is different in the open Serengeti plains. Here the territories are within the home ranges of very large mixed herds (usually ranging from 40 to about 500 members). These mixed herds are composed of females, immatures of both sexes, and fully adult, non-territorial males, and they are, in a sense, in permanent migrations through their often enormous home ranges. Like the female herds in Thomson's gazelle, these mixed herds of Grant's gazelle enter the territories during their daily "circuit." In contrast to the territorial tommy males and the territorial Grant's males on the *mbuga,* the territorial Grant's bucks in the open plains usually tolerate the non-territorial males in their territories. They show a moderate dominance over them, but they do not fight them and do not chase them away–except in very special cases (see p. 152). They show dominance displays toward them and are clearly superior in these encounters. Furthermore, the owner of the territory interferes as soon as one of the non-territorial bucks tries to sexually approach a female. As long as the herd is inside his territory, only the territorial buck has the "right" to court and to copulate. When the mixed herd leaves, he may remain in his territory, or he may leave together with the herd. But, in the latter case, outside his territory he does not behave differently from any non-territorial male until the herd comes back to his territory where he regains dominance over the other males and the exclusive "right" for sexual activities. Thus, in the typical open plains territories of Grant's gazelle, territorial behavior is not combined with harem behavior, but the females pass through the territories, and the non-territorial males are dominated but otherwise tolerated by the owner inside his territory.

This form of territoriality is more or less a compromise with the conditions in the open plains with their large mixed herds. Occasionally a Grant's buck in an open plains territory temporarily behaves more like a territorial male on a *mbuga*. Then, he may chase the non-territorial males out of his territory and/or he may try to prevent the females from leaving it. However, most of these efforts are completely in vain. Even if a territorial buck succeeds in "cleaning" his territory of other males and/or preventing females from leaving it, he cannot maintain this situation for more than half an hour or so. There are simply too many males and females, and when so many females persistently try to leave the (large) territory, or when so many non-territorial males re-enter it again and again from all sides a short time after they have been chased away, even the strongest territorial buck becomes exhausted and gives up. Allowing a certain anthropomorphism, one may say that a territorial Grant's buck in the open plains "would like" to behave like a territorial male on a *mbuga,* but that the situation is hopeless, and, thus, he "resigns" himself to it and "makes the best of it."

9

Rise and Decline of Territoriality

Three phases may be distinguished in territoriality: rise, peak, and decline. The behavior of males at the peak of territoriality has been described in Chapter 7. How far the behavior differs from this peak phase during rise and decline remains to be pointed out. Since in all the species under discussion only adult males become territorial, each of them has to go through a phase of beginning territoriality even in species and/or cases in which the territories are kept for years. Since the non-territorial males are in all-male herds or in mixed herds, it also can be postulated as a general rule that the bucks become territorial from bachelor or mixed groups. Whether there is a phase of decline in territoriality or, perhaps more precisely speaking, how readily such a phase of decline can be observed in a study of a few years, depends very much on the average length of territorial periods in a given species. As pointed out in previous chapters, there are apparently considerable species-specific differences in the average lengths of the territorial periods in Antilopinae. For example, in Thomson's gazelle and Indian blackbuck, most territorial periods last only a few weeks or months, whereas in mountain gazelle and gerenuk, most of them seem to last for several years.

The rise of territoriality has been investigated predominantly in blackbuck, gerenuk, Thomson's and Grant's gazelle. The decline of territoriality has only been studied, to any notable extent, in Thomson's gazelle and blackbuck.

BECOMING TERRITORIAL

Gerenuk

According to Leuthold's (1978a) observations on four individually known young males between one and two years old, these were initially together with an adult female who was known to be the mother of the youngest male, but may also have been the mother of one of the older ones. This female left her original home range one day, but the four young males remained there for a while, forming a bachelor group. During the next year, these bachelors were found on other ranges, but they sometimes returned to their original range. Sometimes all four of them were seen together, sometimes two, sometimes three, or only one. When one of them had separated from the others, he obviously rejoined the group again later. Sometimes one of these males completely disappeared from the observation area for some time. After about one year of movement, splitting, and rejoining, one of the oldest of these bachelors, about three years old, took over a territory in the observation area which had become vacant about two months before. When one of his former companions showed up there, he was chased away. This chased male later disappeared completely from the area, but he returned after about two years and took over a territory which had become vacant three to four months before.

Thus, it emerges that, in this species, the subadult males form small bachelor groups which move through a large area for several years, periodically staying in certain parts of it. During this time, the individual members may temporarily split from the group and later rejoin it. It is not unlikely that a certain search for a vacant territory may take place during this period. When one of such young males reaches adulthood and finds a territory which has become vacant, he takes over and becomes intolerant toward his former companions claiming an area solely for himself. It is particularly noteworthy that such a "new" territory almost totally coincides with the territory of the former holder in this species.

Indian Blackbuck

There are several situations which can lead to becoming territorial in blackbuck. In the simplest case, a solitary wandering adult male happens to enter an area inhabited by females without a territorial buck being present or if territorial males are there, he still has enough suitable space available to establish a new territory. In another case, some subadult or newly mature buck just drops out of a bachelor association as it passes through a favorable stretch and localizes his activity at one place. Such a buck may start a new dung pile by repeated urination and defecation at a particular spot. If there are dung piles from a male formerly territorial at this place, the territorial hopeful will ordinarily use at least some of them. Alternatively, he may establish a new dung pile next to one of the previous owner leaving the former one unused. Such a territorial hopeful may not stay long. He may leave after a while, but he keeps coming back. In a sense, he becomes territorial stepwise.

Except in rare cases where there are no other bucks in the vicinity, the new buck will come into conflict with other males sooner or later. If the place he has chosen is part of a territory which the owner has left temporarily, the established buck, when he comes back a few hours later, will chase away the new male. The new buck may return regularly for several days, but the old owner keeps chasing him away. Two cases of this type were observed. In another case, a subadult male tried to establish himself adjacent to a fully adult, territorial buck. He was fine until they got into a border fight in which the younger male was beaten soundly and chased away. He joined a bachelor herd outside the territorial area, and he never came back to try again. Presumably, he would have remained in possession if he could have proven himself equal in that border fight.

A prominent attribute of cases of ownership change is elevated frequency of dung pile use. Incidence of scraping at dung piles also increases. On peaceful days, well-established bucks may only exhibit actual urination and defecation at a particular dung pile once or twice during any given daylight period. If a newcomer should challenge the established buck's ownership by trying to stay inside his borders, however, the established buck's use of the dung pile may increase by 10 or 12 times. Meanwhile, the newcomer may be using the dung pile frequently himself. If the newcomer becomes less persistent, the established owner's use drops accordingly. Similarly, a buck moving in on an owner who is peacefully giving up his holding shows elevated dung pile use on the order of 10 to 12 times during a daylight period. The former resident may only show use once on his last day before leaving entirely. It takes about two weeks for the new buck to consolidate his position enough for his dung pile use frequency to drop to the one to two times of an established buck.

As fall approaches in Texas, the frequency of aggressive interactions rises and with it the frequency of territorial changes. Once, an adult male in a bachelor herd who had looked poorly during the previous winter, was regaining condition. In late August, he tried to exclude the other bucks from part of the pasture they normally used during their daily round of activities. Some of his displaying was successful, but he ended up fighting one of the bachelors and later a territorial male whose border was not far away. The territorial buck won and chased the other who soon went back to living with the bachelors. During the same period, two newly adult bachelors each succeeded in becoming territorial. One was about three years old and the other about two and a half years old. The elder took a pre-existing territory whose owner had temporarily left. When the latter returned, the new male lost half the territory back to the former owner. Three months later, the younger male became territorial after a fight with another territorial buck.

It must not be inferred, however, that a bachelor who has not been previously territorial in an area simply marches out one day and deliberately starts a fight with a territorial buck. Generally, fights of territorial males with bachelor males are rare. When such a fight takes place, it is usually the bachelor who loses. However, if the bachelor does win a fight against a territorial male, his behavior immediately conforms to the pattern which is typical of a territorial buck.

Winning a fight seems to have a psychological effect linked with the place where the fight occurred. As mentioned above, one territorial male who had no territorial neighbors, used to go to the other end of his pasture and display to the bachelors there if none of them had yet come close enough that day. Since he consistently singled out the most dominant of the bachelors for his target, these two had frequent encounters. Occasionally, this bachelor was slow to withdraw and, in rare instances, returned the territorial male's displays. Over a period of a year, this led to four fights in which the bachelor was always beaten and then chased a long distance. During a fifth fight, the territorial male got a foreleg caught between the other's horns. Although the territorial buck snapped back into position quickly, he soon withdrew from the fight and limped back to his territory. His opponent, who did not press the advantage but chased a nearby female instead, began keeping all the bachelors out of the area where he had won the fight. He held it as a new territory. The territorial male who had lost the fight against him, recovered by the end of the day. However, he stopped using the part of the pasture where he had lost the fight. Also, his excursions in search of bachelors ceased. Instead, he and the new territorial buck displayed to each other from time to time. It should be emphasized that there does not have to be a fight for a buck to become territorial although, in the cases where there is a fight, the question of ownership arises and is decided by the outcome of the encounter.

Under special circumstances, there can be a link between high social rank and territoriality in more than just the general sense of aggressive competitiveness and robust physique predisposing a buck toward becoming a successful territory holder. Two young bucks while making the transition to bachelor status at the same time and joining a small bachelor group with relatively stable membership are likely to develop a fixed dominant-subordinate relationship. Since the group is small and stable, the group also has a fairly stable rank hierarchy. The newcomers become entrenched in this hierarchy as a sub-unit of "superior companion" and "inferior companion." As they mature, the superior companion fills out faster and exhibits aggressive behavior more freely than the inferior companion whom he continually dominates. Eventually, the superior companion seeks to become territorial his first fall as an adult. However, his inferior companion may not make any attempt even after being released from the domineering presence of his superior companion because, by the time the latter has quit the group, other males joining the group in the meantime may have gained ascendency over the former inferior companion. Therefore, his subordinate status within the bachelor group as a whole may be perpetuated. Observations show that a former inferior companion can develop into an aggressive individual capable of holding high rank within his bachelor group and of trying for a territory but only after the bucks above him in the hierarchy have left or have sunk in rank due to some infirmity (Mungall 1978b).

Thomson's Gazelle

In Thomson's gazelle, a buck can become territorial in different ways: (a) during grazing periods while in a mixed herd (mainly in the beginning of the wet season), (b) by the dissolution of a group of adult bachelors into a territorial mosaic, (c) by solitary (or with only a few companions) wandering around "in search" of a territory, (d) by occupying part of the formerly mutual home range of a bachelor group, or (e) by conquering (back) an occupied territory. There may be more ways of becoming territorial in Thomson's gazelle, but these were the ones actually observed.

As stated several times in previous chapters, the typical habitat of the Thomson's gazelle is the open, short-grass plains. However, during the dry season, they retreat more and more into forests. Toward the end of the dry season, green food plants also become rare in the forest areas, and/or the feeding conditions there may become poor due to extensive burning. As soon as the first rains fall in the open plains, the gazelles return to this preferred habitat where short green grass is sprouting rapidly. Mixed herds are the typical formations during this migration, and they move in long files. When these herds arrive in an open area with favorable feeding conditions, the animals naturally begin to graze intensively with interruptions only for resting and shorter moves. Predictably, the adult bucks move to the periphery of a herd during such long grazing periods, and then, typically, the long file of females, immatures, and juveniles is surrounded by a chain of grazing adult males. This formation is not an anti-predator behavior since tommy males neither defend themselves nor the females and the young against predators. They also are not more alert than females; rather, it is the opposite. Generally, the individual distances which adult bucks keep during grazing are considerably greater than in any other activity and are considerably greater than the females' distances (Walther 1977a). Thus, the described grazing formation simply results from the large individual distances of grazing adult males. During grazing, these males move farther and farther away from the herd, and enlarge the distances between each other, spreading like a fan, far beyond the distances they usually keep when grazing in other situations. In spite of these enlarged individual distances, more hostilities occur between these males. For example, one of them may cease grazing and march to the next one, easily over a distance of 50 to 100 m, in order to threaten or to fight him. Some of the males may retreat to the herd after such encounters, but most of them keep their positions. Also, the marking activities (marking with preorbital gland secretion as well as marking by urine and feces) increase during these long grazing periods. After a few days, sometimes even after a few hours, these males have established territories and territorial boundaries. In such a case, several–sometimes many–bucks become territorial at the same time forming a territorial mosaic. Thus, they become territorial while grazing, starting out with the normal grazing distances between adult males which bring them to the periphery of the mixed herd. These grazing distances are enlarged more and more until they finally become equivalent to territorial distances (usually about 100 to 300 m from the center of one territory to the center of the next one). For a short time, the male is the center of this enlarged individual distance, and it moves with him. But soon he "fixes" the enlarged individual distance on the landscape and it becomes a territory with a "geographical" center, i.e., the position of the center as well as that of the boundary do not move or change anymore with the movements and the position of the male.

In a similar way, several bucks may simultaneously become territorial from an all-male group, although grazing does not play such a prominent role in this case. This happens predominantly when a bachelor group is in the range of up to 30 males and largely, or even exclusively, composed of adult bucks, which is by no means rare in Thomson's and Grant's gazelles (Walther 1972a) or in blackbuck. Then, in the course of a few days, sometimes even within a few hours, these bachelors may enlarge the distances between each other, finally ending up with territories. In this situation, as well as in becoming territorial from a mixed migratory herd, the first fights among the prospective neighbors can be very violent. It can happen that one of the combatants is defeated completely in such a fight, flees and is chased by the victor over a long distance, and loses his territorial status in this way before it is fully established.

Furthermore, a male may leave a mixed herd, or, probably more often, a bachelor group and wander around solitary with a few (one to four) companions, presumably searching for a territory. Often such males return to the all-male herd without having tried to become territorial. For this reason, one may have reservations whether becoming territorial is always the purpose of this solitary wandering. However, when such a solitary wandering buck becomes territorial, he usually establishes his territory at the periphery of a territorial mosaic. Sometimes he may also find an abandoned territory within such a mosaic. If a few companions originally were with him, he now chases them away. They may establish territories later, or they may re-join an all-male group or mixed herd.

Sometimes only one male in a bachelor herd becomes territorial, claiming part of the formerly mutual home range exclusively for himself. It often takes days before the other males "realize" this new situation. They freely and frequently enter such a territory as they were accustomed to do when it was part of their home range. Thus, the establishment of such a territory needs an unusually great amount of threatening, fighting, and chasing on the part of the territorial hopeful. Not rarely, he gives in and becomes a "normal" member of the bachelor group again after a short while. In cases in which such a buck was eventually successful, he always was the strongest and the most dominant male in the former group. These were the only cases in which a direct relation between territoriality and high rank in a social group was noted in this species.

Finally, a bachelor may occasionally conquer an occupied territory. Apparently, laymen often presume that this would be the usual way in which a male becomes territorial. Therefore, it is perhaps not superfluous to emphasize that this is by far the rarest case of all in species such as Thomson's gazelle. Usually, a bachelor does not fight a territorial male but withdraws or even flees when charged. When a bachelor occasionally takes the challenge and fights back, he usually is defeated. In one of the exceptional cases in which a bachelor was observed to keep up in fights against a territorial male, this individual was well-known from previous observations. He had been territorial on this location but had left during a migration period. When he returned months later, he found part of his former territory occupied by another male. Thus, this "bachelor" fought for re-occupation of his former territory. In the other few cases in which bachelors were observed to be successful in encounters with territorial bucks, the "history" of the individuals was unknown. It is very possible that it was similar to that of the buck known individually.

In species such as Thomson's gazelle, Grant's gazelle, and Indian blackbuck where the males more or less regularly leave their territories after some time, the question arises whether they have to establish new territories later in the same or other areas. In some cases, the return of a male to a formerly held territory after a period of absence has been observed in all three species. In a few other cases, males returned to the area in which they had been territorial before, and did establish territories there, but only in the vicinity, i.e., not precisely at the same places as their former territories. We do not know whether, in his later life, a male becomes territorial only at or near places where he has been territorial before, and/or whether he may establish another territory during the time of absence from a given (observation) area.

Grant's Gazelle

In Grant's gazelle, only two ways of becoming territorial have actually been observed: (a) establishment of territories by solitary wandering males who had split from bachelor groups or mixed herds and were wandering alone or with one or two companions, (b) establishment of a territory by claiming part of the home range of a bachelor group which the territorial hopeful had previously used together with the other males. There is no principal difference from the corresponding cases described in Thomson's gazelle. However, quite frequently an additional factor seems to be involved in Grant's gazelle which is not evident in tommy. It happens that an adult Grant's buck starts becoming territorial at the moment when he meets a single female who has a (lying out) neonate fawn.

DECLINE OF TERRITORIALITY

As previously mentioned, the decline of territoriality was studied mainly in Thomson's gazelle and blackbuck. When a tommy male continues to stay in his territory for more than three to five months, the local conditions often have deteriorated drastically. For example, he may have established his territory during the wet season when the area was covered with short green grass, many females and bachelors were around, and his territory was surrounded by neighboring territories. Beginning with the onset of the dry season, the country dries up, the grass becomes yellow or grayish, the originally short grass may grow high, and the majority of females and bachelors, and even many of the territorial males around him leave the area. Now, such a buck is often alone in his territory for days and weeks. If one of the territorial neighbors is still present, they may engage in encounters over the boundary now and then. His marking activities dwindle. If some females occasionally happen to pass through the area and enter his territory, the buck herds them for a short while, but he usually does not make great efforts to prevent them from leaving. If bachelors enter his territory, the territorial buck is dominant over them, but he does not chase them away. Often they stay in his territory for hours. When the females or the bachelors move ahead, the territorial buck may sometimes leave his territory with them. After a mile or so, he may stop and return. Even when alone, he sometimes makes excursions outside his territory. Altogether, it seems that he is not quite sure about his territorial status and the position of the territorial boundaries anymore during this decline phase, and sooner or later he disappears.

Local conditions also vary for blackbuck, but not as drastically as for tommy males. Blackbuck females and bachelors are frequently near since blackbuck do not migrate on the Texas ranches and in the sanctuaries as they are now in India. Bucks in prime territories seldom lack neighbors since at least some males are territorial during all seasons. Grass in the characteristically heavily grazed Texas pastures tends always to be too short to sustain a territorial buck without daily feeding trips outside his domain. Similarly, many Indian bucks in heavily used areas or with extremely small territories probably never have enough grass within territory borders to support themselves. But in spite of the more constant conditions, the rigor with which a blackbuck male keeps his territory still wanes eventually.

After several months, the buck may spend more time away feeding and may become slower to return after fleeing a disturbance. Finally he may only show up at dusk and may leave again early in the morning with hardly a brief visit back in between. He may allow bachelors to remain in his territory for hours provided they create no disturbances, and he may become lax about challenging neighbors approaching his boundary. Bucks whose tenancy approaches a year or more often look thinner near the end of their territorial period, and poorer summer and winter grazing probably aggravate their decline. However, even bucks territorial for lesser periods who leave during spring or fall show the same symptoms of decline in territorial behavior. A few weeks after leaving or losing a well-established territory, a buck may seek to hold a territory elsewhere but he will abandon this, too, after only a brief stay. One such buck made short tries in one and later a second other location.

ABANDONING THE TERRITORY

Not all territorial males go through the three phases (beginning, peak, and decline) of territoriality described above. It has already been mentioned that, at least in some species such as Thomson's gazelle, males sometimes try to become territorial but leave already during this initial phase, without having established a truly "good" territory. Other males leave their territories right in the middle of that phase which was described as the peak of territoriality and without having displayed any signs of declining territoriality. That a territorial male loses his territory to another buck, is rather exceptional as stated above. Thus in most of the cases, a male just quits being territorial one day.

Some people are surprised when they hear this and find it hard to understand. However, when one has observed a buck in a very "active" territory–chasing the bachelors, herding and courting the females, fighting the neighbors, marking, etc., all day long and into the night–it is very understandable that such a male may eventually become exhausted and may simply give up his territorial status one day. Furthermore, one must not forget that all these species are basically gregarious animals, and, thus, the tendency for territorial isolation may easily run into conflict with the tendency to keep company with conspecifics and to participate in their activities and movements. This inner conflict may come into play particularly in those cases in which a territorial individual has no, or only little, opportunity for interactions with conspecifics. It may eventually cause him to abandon a non-frequented territory in order to satisfy his social needs. Finally, the suspicion often could not be avoided that there is some kind of an action-specific potential for territoriality in these animals, as ethologists have generally postulated for all kinds of instinctive behavior, and that this action-specific energy can become consumed under certain conditions. Three events from the life of Thomson's gazelle strongly suggest the exhaustion of such a "territorial potential." Two of them are already familiar to the reader. When a territory is "flooded" by migratory mixed herds for hours and days, the owner may first show a true outburst of chasing the males and herding the females, but eventually he gives up, moves with them and abandons his territory (p. 51). The second case is when the majority of the gazelle have left an area in the beginning of the dry season, but a few territorial males may have remained there. When, after weeks or months, a mixed herd happens to pass through this area one day, it "collects" the territorial males in passing. They leave their territories and go with the herd (p. 51).

Perhaps the most instructive example was provided by the study of the tommy buck with the longest uninterrupted territorial period recorded (11½ months). This male, named "Short-Tail," became territorial during the short rains, and kept his territory throughout the long rains and the dry season into the beginning of the next short rains. In the second half of the dry season, he showed all the signs of declining territoriality in an almost classic way. With the beginning of the next rainy season, the tommy herds returned to the area in huge numbers. Females and bachelors frequented "Short-Tail's" territory every day, and he was again surrounded by territorial neighbors as when he established his territory almost one year before. For about three weeks, his declining territoriality came back to full life, and he behaved again as any "good" territorial buck at the peak of a territorial period. Then, he abandoned his territory without any recognizable reason from one day to the next, while fresh, green grass was abundant and many females were roaming the area, etc., in short, under conditions which were ideal for a territorial tommy male. It looked as if his "territorial potential" had dwindled down in the long period of territorial decline during the drought, and this remaining potential had been completely consumed within a few weeks of full territorial activity after the environmental and social conditions had changed for the better.

10

Functions of Territoriality in Antilopinae

FUNCTIONS IN REPRODUCTION

When discussing the functions of territoriality in Antilopinae, it is perhaps not out of place to point out first those functions which territoriality does *not* serve in these animals. There is a lot of theorizing and hypothesizing in the recent literature about the biological functions of territorial behavior. Usually, such speculations try to get across the point that territoriality would serve to ensure an optimal food supply to the territorial animals (including their mates and offspring), that potential food competitors would be excluded from the occupied area, and/or that the territorial spacing-out would benefit the species in that it would allow a more economical use of the available food. These and similar hypotheses are largely invalid for gazelles and Indian blackbuck except possibly for the gerenuk whose territoriality seems to deviate somewhat from territorial behavior as it occurs in the majority of Antilopinae species (see p. 193).

Food, feeding behavior, feeding style, etc., are of minor importance for the territoriality of the species under discussion. Of course, it is not a disadvantage when feeding conditions are good in a territory. Also, feeding conditions in a given area may indirectly favor or disfavor the establishment and the maintenance of a territory in that the conspecifics, particularly females, readily come to and stay in an area with good feeding conditions, whereas they leave when the food becomes scarce. On the other hand, the owner can tolerate poor feeding conditions to a remarkable extent, particularly after a territory is well established. On some Texas ranches where the animals depend on supplementary food from feeders, Indian blackbuck males routinely leave their territories in order to feed–inasmuch as the feeders are not inside their territories–but then return to their territories and stay there. Actually, this pattern of daily absences for feeding is common even for blackbuck subsisting on native vegetation. In heavily grazed Texas pastures where grass may only be as tall as 3 cm for long periods, territories are still frequently only 1 to 4 ha in size. Nevertheless, just such an area yielded one of the longest territorial periods observed (more than 11 months) even though the owner looked noticeably poorer in condition during the end of his stay. Similarly, the Indian bucks on the dense territorial mosaic at Velavadar can hardly be expected to maintain themselves through the summer dry season on the scattered sprigs of 2 to 6 cm high grass within holdings of less than half a hectare.

A territorial tommy buck carefully keeps his–at maximum–four to six neighbours out of his territory but readily allows 100 or more females to come into his territory and to graze there for hours. He even tries hard to prevent them from leaving the territory. Thus, one can hardly argue that the male's territoriality would serve to ensure the food supply for him. Likewise, it is hard to see any indications of a more economic use of the food sources in species where proportionally only a few (male) animals are spaced out and stay in limited places due to their territoriality, while hundreds and thousands of non-territorial conspecifics, particularly females, roam and graze in the entire area.

Also, the hypothesis that the territorial male provides a food resource and optimal feeding conditions for "his genes" in the growing offspring, is completely out of place. The minimum gestation period in the Antilopinae species is about five and a half months. (In many of the species, it is longer.) Under natural conditions, the females do not stay in a territory for such a long time. Thus, a female usually gives birth in another area which is often far away from the territory where she was bred. Moveover, in some of the species, most males do not keep their territories long enough for the females to give birth to that male's offspring while in his territory.

In previous chapters, we mentioned several times that territorial behavior is linked to reproductive behavior in Antilopinae, and reproduction obviously is its most important function in these animals. Whether it is the only function is a difficult question. The fact that Antilopinae males can become territorial without females present, and that at least some of the males keep their territories for considerable lengths of time after the females have left the area and/or during seasons with low or no reproductive behavior, may indicate that territoriality probably serves more biological functions than reproduction in these species. However, these functions are not clear at present.

Obviously, the territorial isolation of the Antilopinae males makes it possible to overcome certain negative consequences of living in herds. As mentioned earlier (p. 90), non-territorial adult males sometimes sexually approach females in mixed herds or when a bachelor group occasionally meets a herd of females. Such events were observed in Thomson's gazelle, Grant's gazelle, mountain gazelle, and Indian blackbuck. Then usually several males run after the same female. They release flight and/or hiding reactions in her, and they almost regularly start fighting each other, etc. Not one case of successful mating has been observed in such a situation in all our studies. Theoretically, copulation does not appear to be impossible in such a situation, but it certainly is a rare exception under natural conditions. On the other hand, in a territory, there either are no other males, or they are so subordinate to the owner that they do not "dare" sexually drive a female in his presence. Thus, the owner does not have to interact with competitors during courtship, and he can focus his efforts on the female. Since only one adult buck is approaching her, the probability that the female may flee from him is more reduced than in the case of several males. If she flees, a territorial buck will try to place himself between her and the boundary, and there is the chance that he may stop her. Non-territorial males in the herds may also chase a fleeing female, but since there are no territorial boundaries, they do not try to pass her and herd her back. Thus, she runs ahead unobstructedly with increasing speed. In short, territoriality facilitates successful mating by counterbalancing the negative effects of gregariousness in the gazelles and their relatives.

It can happen that a male courts a female and copulates with her when he is just in the process of establishing a territory. Thus, successful mating can take place even in a territory which is held only for a few days or even a few hours. However, a buck truly becomes a factor in the reproduction of the population only when he stays for a longer time in his territory. Provided that females frequent his territory more or less regularly, and that a good number of them consecutively come into heat, a male in a well-established territory may breed approximately one female per day over a period of weeks or months.

FUNCTIONS IN SOCIAL ORGANIZATION AND SPATIAL DISTRIBUTION

As mentioned above, territoriality may serve functions other than reproduction in Antilopinae, but they are not as clear or only appear to be more or less side effects. One of these functions is the antagonism of territoriality to migratory behavior; the territorial males are the "brakes of migration" (see p. 51). Furthermore, territorial bucks often chase non-territorial males out of the mixed herds when they pass through their territories. In this way, they contribute to the separation of the bachelors from the females. Apparently, some authors (e.g., Jarman 1974) are inclined to explain the existence of all-male groups by this activity of the territorial males. At least in its pure form, this hypothesis must be rejected because (a) all-male groups also occur in bovid species (e.g., wild sheep and goats) in which there is no territorial behavior, and (b) in species, such as Antilopinae, where some males become territorial and charge trespassing bachelors, the latter frequently are chased in completly different directions. Thus, the chasing behavior of the territorial males contributes to the separation of non-territorial males from females, but the (chased) bachelors' coming together and forming all-male groups cannot be explained by these activities of the territorial bucks.

On the other hand, the activities of the territorial males may have an influence on the spatial distribution of the bachelor groups in a given area. This influence is particularly true for species in which the territories form a relatively dense mosaic. The female herds can roam through the entire area, i.e., the area occupied by the territorial males as well as the unoccupied surroundings, but bachelor herds are excluded from the territorial area. Thus, they usually stay at the periphery of a locality inhabited by gazelles. This does not necessarily imply that the bachelor herds would be forced to stay in areas with poor feeding conditions. At least in a natural situation as in the Serengeti plains, there is hardly any evidence for such an assumption. However, the all-male groups' position in peripheral areas has another disadvantage. Since predators frequently approach from the periphery, they meet the bachelor herds first, and, thus, the non-territorial males may have to pay a particularly great toll to predation.

FUNCTIONS IN INTERSPECIFIC AND PREDATOR-PREY RELATIONSHIPS

Under natural conditions, territorial competition is restricted to conspecifics in Antilopinae and other bovids. Consequently, animals of other species can freely enter and pass through the territories. When the males of another species also are territorial, their territories may overlap the territories of the Antilopinae males either completely or in part. Under certain conditions, such as when two territorial males of different species (e.g., Thomson's gazelle and topi, etc.) have occupied the same place and both of them have little interaction with conspecifics for a long time, these males may even establish an absolutely positive relationship with each other. They synchronize their activities and keep close to each other while grazing, moving, standing, and resting. The question of whether such a relationship between territorial males of different species may have a favorable effect (e.g., with respect to a continued stay) on their territoriality, cannot be answered at present.

In captivity and sometimes also in such a semi-natural situation as in blackbuck in Texas pastures, territoriality can have more effects in interspecific relationships than under natural conditions. In captivity, Antilopinae may sometimes treat animals of other (ungulate) species like conspecifics, quite according to Hediger's (1942) "assimilation tendency." Of course, this situation also includes the possibility of territorial competition between animals of different species. The territoriality of the Antilopinae males usually has no great effects upon animals of another species in such cases because most of the ungulate species kept together with them are superior to them in size and strength. Thus, the territorial Antilopinae bucks may charge them but usually are beaten. Such defeats may have negative consequences on the Antilopinae males, and, in general, the territorial behavior of males of other, bigger ungulate species may negatively influence the territorial behavior of Antilopinae under captive conditions.

As previously mentioned, no indications were found that territorial Antilopinae males were more vulnerable to natural predation than conspecifics from other social classes. An indirect function of territorial behavior in predator-prey relationships may be seen in those cases in which the territorial bucks force the non-territorial males to stay in peripheral areas where they are particularly exposed to attacks by predators, as mentioned above. In special situations, another (minor) function of territorial behavior in predator-prey relationships was observed. When predators such as lions come to an area occupied by territorial tommy males during the daytime, they often do not have immediate hunting intentions. If there are females in the territories, they usually leave the neighborhood of the predators sooner or later, but the territorial bucks remain in their territories and keep the predators under close surveillance. As long as there is a predator in the vicinity of the territory, the owner watches him uninterruptedly (Figure 72). If the predator moves off, the territorial buck follows but remains inside his territory and at a distance from the predator.

When the buck comes to the boundary of his territory, his territorial neighbor becomes alert. Often the latter may not have seen the predator yet, but he reacts to the neighbor's approach. In running toward the boundary, he becomes aware of the predator and now he "takes care" of it as long as it is close to his territory. This may be repeated with the next neighbor, etc., until evening when the predator begins actively to hunt. By then, its position is known to the tommies in the surroundings due to the behavior of the territorial males, and, thus, the predator's chances of a surprise hunt are gone, or at least diminished.

Figure 72: Surveillance of a predator by a territorial tommy male. (a) A lioness has approached a tommy territory at noon and has lain down in the shade of a tree. The territorial buck approaches the lioness. (b) The territorial buck has bedded down facing the lioness and keeps her under uninterrupted watch as long as she is in the vicinity of his territory. (Photos: F.R. Walther–Serengeti National Park, Tanzania.)

In such special cases, one may speak of a function of territoriality in predator-prey relationships. However, the tendency to keep predators under observation is by no means restricted to territorial males. It is also found in females and non-territorial males. Thus, in principle, this behavior is not a special anti-predatory device of the territorial individuals.

11

Comparative Aspects of Antilopinae Territoriality

COMPARISON WITHIN THE ANTILOPINAE SPECIES

Before we compare the territorial behavior of Antilopinae to that of other bovid species, it seems advisable to compare the phenomena of territoriality among the Antilopinae species themselves (Table 9). A short summary of the previous chapters will provide this comparison.

In more than one regard, the territorial behavior of the gerenuk deviates most from the picture of territoriality in other Antilopinae species investigated to date. For one, open plains and short-grass areas are the localities on which the other species prefer to establish their territories, whereas gerenuk live in comparatively densely vegetated "bush" country. Secondly, gerenuk is the least gregarious among the Antilopinae species. In Leuthold's (1978a) observation area in the Tsavo National Park, mean group sizes, somewhat varying with sex and season, ranged between 2.14 and 4.64, and the maximum group size recorded was 13 animals. These sizes are nothing compared to most of the other Antilopinae species, provided that their populations have not been drastically reduced by excessive human hunting. Since territoriality is susceptible to social and environmental conditions, some deviations from the usual picture have to be expected already on the basis of these two factors in the case of the gerenuk. Apparently Leuthold (1978a), who certainly is an expert on territoriality in African bovids, is not even quite sure whether he should speak of territorial or home range behavior in gerenuk (see p. 69). Based on the observation that the home ranges simultaneously inhabited by adult gerenuk males never overlapped, Leuthold finally states that "their home ranges . . . can be termed territories."

Other major differences in the territorial behavior of gerenuk as compared to other Antilopinae species seem to be that (a) all the adult gerenuk males in a given area (at least, in Leuthold's study area in Tsavo Park) become territorial, and consequently, the bachelor groups consist exclusively of adolescent and

Table 9: Territoriality and Territorial Marking in Antilopinae Species Investigated to Date

Species	♂ territorial. ♀♀-herds temporarily in territory.	♂ territorial. Stable harem in territory.	Dung piles in territory.	Preorbital gland secretion marks in territory.	Remarks
Antilope cervicapra Indian blackbuck	++	–	++	+	
Antidorcas marsupialis springbok	++		+	–	
Ammodorcas clarkei dibatag	?		+	?	
Litocranius walleri gerenuk	+	?	+	+	In part after Leuthold (1978).
Gazella (N.) granti Grant's gazelle	++	+	+	–	Locally mixed herds instead of ♀♀-herds temporarily in territory.
Gazella (N.) soemmeringi Sömmering's gazelle	?		?	–	
Gazella (G.) gazella mountain gazelle	+	–	++	–/+	Secretion marking in Indian subspecies *(bennetti)*.
Gazella (G.) rufifrons red-fronted gazelle	?		+	+	
Gazella (G.) thomsoni Thomson's gazelle	++	–	++	++	
Gazella (G.) dorcas dorcas gazelle	?		+	–	
Gazella (T.) subgutturosa goitered gazelle	?		++	++	

Signs and abbreviations: + = clearly observed
++ = very pronounced and/or frequent
? = probably present but not beyond possible doubt
– = lacking
empty = unknown at present
(N.) = subgenus *Nanger*
(G.) = subgenus *Gazella*
(T.) = subgenus *Trachelocele*

subadult males, and (b) the territorial bucks hardly ever interact with each other in agonistic encounters over the boundary (no observations during the three years of Leuthold's study). As a minor difference, one may list that when gerenuk males reach adulthood, they always seem to become territorial by wandering around in a given area and eventually finding a suitable, unoccupied locality, whereas in other Antilopinae, this "searching for a territory" is only one way among others to become territorial. Chasing of bachelors by territorial males has been recorded in gerenuk, but Leuthold (1978a) does not mention the herding of females which is a frequent and striking activity in the territorial males of the other Antilopinae species. Possibly, the gerenuk buck's marking of females with the secretion of his preorbital glands, a behavior shared only by the dibatag male, may "substitute" for herding, in a sense. Furthermore, gerenuk females appear to stay with a buck when their home range more or less coincides with his territory. Thus, territorial herding may not be as necessary in gerenuk as in other species. Finally, the size of gerenuk territories (130 to 340 ha) even somewhat exceeds the territory size in mountain gazelle (100 to 220 ha), as the species with by far the largest territories among the other Antilopinae.

In all the other Antilopinae species investigated, the similarities of territorial behavior are much more prominent than the differences, as is evident from our previous descriptions and discussions. That territorial behavior can be combined with harem behavior in Grant's gazelle does not appear to differ too drastically from the situation in other species. For one, in Grant's gazelle, harem behavior only occurs under certain conditions. Secondly, the possibility that such a combination of territorial and harem behavior may occasionally also occur in other Antilopinae species under special conditions cannot be excluded, although it has not yet been observed to date. Similarly, territorial Grant's bucks' occasional tolerance of bachelors within their territories does not necessarily imply a principal difference. At least, it has a parallel in the tolerance of territorial tommy males toward bachelors during the decline phase of territoriality and in the occasional tolerance of territorial blackbuck males toward bachelors when no females are present. Thus, it is more a gradual difference, i.e., it is more frequent and it is due to a specific situation in Grant's gazelle. In principle, however, it can also occur in other species.

The step-wise progress of courtship as described in mountain gazelle where a courtship ritual may last more than one day, certainly is different from other species where a mating ritual lasts a few hours at maximum, and most of the rituals take less than one hour. On the other hand, this difference is more in sexual behavior than in territorial behavior. It may be somewhat related to the size of the territory in that a lengthy mating ritual of the Antilopinae type, i.e., the pair moving during courtship, requires a comparatively large territory. If, in a species with small territories such as tommy, the mating ritual would take several hours or even days, the female very likely would have left the territory before the buck is half through with his courtship.

Differences in the marking of the territory by depositing secretion from the preorbital glands are evident within the Antilopinae species since not all of them mark objects (p. 87). In combination with relatively small territory size and great density of neighboring territories, the object marking may account for the observation that such species appear to be more independent of environmental landmarks when establishing and keeping their territories than species which do not mark objects and have larger territories with greater distances between them.

These considerations lead us to the differences in size and density of territories. Even when one excludes the somewhat aberrant case of the gerenuk from the discussion, the differences in territory size are still enormous, ranging from 100 to 220 ha in mountain gazelle, over 15 to 50 ha in Grant's gazelle, down to 1 to 20 ha with an average of about 4 ha (exceptionally even only 0.3 ha) in blackbuck, and 1 to 10 ha with an average of 3 ha in Thomson's gazelle. Of course, territory size may vary with suitable space availability and with population density; however, it obviously varies only within species-specific limits. Within these species-specific limits, one may assume a tendency to make the territory as large as circumstances allow. However, even when a tommy buck, for example, is alone in a suitable area, the size of his territory will never approximate a quarter of the territory size which apparently is quite common in mountain gazelle. The size of the territories, of course, is related to their density. A truly dense territorial mosaic can only occur with small territories. Due to the differences in territory size and density in the different Antilopinae species, territoriality has two opposite effects with respect to the frequency of agonistic encounters among territorial bucks. Large territory size diminishes the frequency of interactions between neighbors, and, of course, this diminution becomes even greater when broad strips of "no-man's land" separate the territories from each other. With immediately adjacent territories and small territory size, the opposite is true.

Regardless of whether the agonistic encounters among territorial bucks are rare or frequent, territorial neighbors are peers, and in agonistic encounters among peers the probabilty of fighting is always greater than in agonistic encounters among unequal opponents. Furthermore, aggressiveness undergoes a certain ontogenetic development in males. Immature males are highly aggressive in that they have many interactions with each other and a great proportion of these consist of fighting. Our material does not allow a conclusion that agonistic encounters are less frequent in non-territorial adult males, probably they are not, but our data in all the species investigated show that when the males reach adulthood, threat and dominance displays become more elaborate and frequent, and they replace fighting in many cases. Thus, the proportion of fights in agonistic encounters of non-territorial adult males with each other is reduced as compared to the proportion of fights in the agonistic encounters of subadult and adolescent males. In this sense, one may say that non-territorial adult males are less aggressive than immatures. When a male becomes territorial, his aggressiveness increases as is expressed by his intolerance toward bachelors and by the great proportion of fights in the encounters with territorial neighbors. In this regard, the territorial bucks behave similarly to subadults. In contrast to the immature males, however, the encounters among territorial peers are almost regularly initiated by intensive and often protracted threat and dominance displays and/or space-claim displays, the fights are often brief, and the forms of fighting or the fighting techniques are more limited than in the fights of the young males, and, finally, there usually is no winner and no loser in the territorial fights.

These ritualized interactions among the neighbors seem to have positive effects on territoriality. Firstly, they contribute to making the boundary line more precise, and the determination of precise boundaries appears to be generally of advantage to the territorial status of an individual. Secondly, the neighbors give a certain "hold" to each other by these "dear enemy" relationships, i.e., the more frequent such interactions are, the more the territorial bucks become independent of environmental factors, particularly of landmarks.

Finally, there are interspecific differences with respect to the length of the territorial periods in Antilopinae. The extremes are exemplified by mountain gazelle and gerenuk on one hand, and by Thomson's gazelle on the other. In gerenuk, territorial periods between one and three years, and in mountain gazelle, territorial periods of approximately two years, were recorded. At least in the latter species, these were only fractions rather than complete territorial periods. It does not appear too far-fetched to consider the possibility that a male may establish a territory only once and then keep it for life provided conditions remain favorable. In contrast, the longest uninterrupted stay of a tommy male recorded was not quite one year, and most of the periods ranged between one to five months with a clear bias toward the shorter periods.

One could perhaps say that there is a general tendency to keep well-established territories as long as possible even in the species with comparatively short territorial periods. One may even speculate that they, too, would keep their territories all year long (a) if there were no marked seasonal changes in the environmental conditions (climate, food, water) which more or less necessitate considerable movements and may possibly induce internal (endocrine?) changes in the animal, (b) if the situation in the population (density, composition and proportions of males and females, etc.) were stable in a given area, and, partly linked with this last point, (c) if the activities of the territorial males (particularly herding and courting females, chasing bachelors, interacting with territorial neighbors) could be so delicately balanced in frequency and intensity that they neither are used too infrequently resulting in a decline of territoriality nor too frequently resulting in a consumption of the "territorial potential." In species such as Thomson's gazelle, the fulfillment of all three of these points is unlikely under natural conditions. However, in captivity (as in a good zoological garden) at least the conditions of less drastic changes in the environmental conditions and of a more stable population situation may be fulfilled. Thus, it appears to be possible that the territorial periods sometimes last longer under captive conditions than in the wild in such a species.

It is not easy to speculate on the phylogenetic evolution of territoriality in Antilopinae because territoriality is composed of many behavioral features, and it often is not certain which of these features should be considered with respect to the evolution of territoriality. Also, when one chooses some of these features, it is often not clear whether they indicate an advanced or a primitive state. For example, when one is inclined to consider increasing differentiation of territories from home ranges, increasing determination of the territorial boundaries, and increasing independence of the establishment of territories from environmental factors, such as landmarks, to be criteria of an advancement in evolution, one might come to state that among the species investigated, gerenuk shows features of a primitive form of territoriality, that mountain gazelle, Grant's gazelle, and springbok–apparently in this sequence–show indications of an increasing advancement in territorial behavior, and that, finally, Thomson's gazelle and blackbuck seem to represent the most advanced stage among the Antilopinae. However, one may come to a different ranking when using other criteria.

COMPARISON WITH OTHER BOVID SPECIES OF THE TERRITORIAL/GREGARIOUS TYPE

In quite a number of the about 100 recent species of horned ungulates (Bovidae), it is still unknown whether they are territorial or not (Table 10). Of the species in which our present knowledge allows us to make a definite statement, roughly 50% have been found not to be territorial. These are mainly the wild oxen (Bovinae), the spiral-horned antelopes (Tragelaphinae–or Tragelaphini according to other classification systems), the wild sheep and goats (Caprinae) including some distantly related species such as mountain goat *(Oreamnos americanus)*. Other than Antilopinae, territorial behavior has been found mainly in the dwarf antelopes (Neotraginae–or Neotragini, respectively), in wildebeest, hartebeest, topi and related species (Alcelaphinae), and in the waterbuck group (Reduncinae). Furthermore, it also seems to exist in at least some of the oryx species and their relatives (Hippotraginae) and it is likely in the duikers (Cephalophinae) although to date none of the numerous duiker species has been thoroughly investigated in the wild with respect to territorial behavior.

Table 10: Occurrence and Types of Territoriality in Bovidae Genera

Subfamily*	Genus	Terr.	♂/♀	♂ + ♀	♂ (+ ♀♀)
Cephalophinae	*Cephalophus*** duikers	?	?	?	
	*Sylvicapra*** grey duiker	?			
Neotraginae	*Neotragus*** royal antelope				
	*Nesotragus*** sunis				
	*Madoqua*** dikdiks	?		?	
	*Rhynchotragus*** trunked dikdiks	+		+	-
	Oreotragus klipspringer	+		+	-
	Dorcatragus beira				
	*Raphicerus*** steenbok	+	+		-
	*Nototragus*** grysbok				
	Ourebia oribi	+			?
Antilopinae	*Antilope* blackbuck	+	-	-	+
	Antidorcas springbok	+	-	-	+
	Procapra Mongolian gazelles				
	*Gazella*** small gazelles	+	-	-	+
	*Nanger*** larger gazelles	+	-	-	+
	*Trachelocele*** goitered gazelle	+			?
	Ammodorcas dibatag	?			
	Litocranius gerenuk	+	-		+

Table 10: (continued)

Subfamily*	Genus	Terr.	♂/♀	♂+♀	♂(+♀♀)
Aepycerotinae	*Aepyceros* impala	+	–	–	+
Saiginae	*Saiga* saiga				
	Pantholops Tibetan antelope	?			?
Reduncinae	*Kobus*** waterbuck	+	–	–	+
	*Adenota*** kobs	+	–	–	+
	*Hydrotragus*** lechwes	?	–	–	?
	*Onotragus*** Nile lechwe	?			
	Redunca reedbucks	+	–	?	?
	Pelea rhebuck				
Hippotraginae	*Oryx* oryx antelopes	?			?
	Addax addax antelope				
	*Hippotragus*** roan antelope	–	(–)	(–)	(–)
	*Ozanna*** sable antelope	?			?
Alcelaphinae	*Damaliscus* topis	+	–	–	+
	Alcelaphus hartebeests	+	–	–	+
	Connochaetes wildebeests	+	–	–	+
Nemorhaedinae	*Nemorhaedus* goral				
	Capricornis serow				
Budorcatinae	*Budorcas* takin				
Ovibovinae	*Ovibos* musk ox	–	(–)	(–)	(–)
Rupicaprinae	*Rupicapra* chamois	?			?
	Oreamnos mountain goat	–	(–)	(–)	(–)
Caprinae	*Hemitragus* tahr	–	(–)	(–)	(–)
	Ammotragus aoudad	–	(–)	(–)	(–)
	Pseudois blue sheep	–	(–)	(–)	(–)
	Capra goats	–	(–)	(–)	(–)
	Ovis sheep	–	(–)	(–)	(–)

Table 10: (continued)

Subfamily*	Genus	Terr.	♂/♀	♂+♀	♂(+♀♀)
Tragelaphinae	*Tragelaphus* spiral-horned antelopes	–	(–)	(–)	(–)
	Taurotragus eland antelopes	–	(–)	(–)	(–)
	Boselaphus nilgai antelope	–	(–)	(–)	(–)
	Tetracerus four-horned antelope				
Bovinae	*Anoa*** anoa				
	*Bubalus*** water buffalo	?			
	Syncerus African buffalo	–	(–)	(–)	(–)
	*Bibos*** gaurs	–	(–)	(–)	(–)
	*Novibos*** kouprey				
	*Bos*** aurochs***	–***	(–)	(–)	(–)
	*Poephagus*** yak				
	Bison bison		(–)	(–)	(–)

* Some of these subfamilies are considered as tribes by certain taxonomists.
** Subgenera which are treated equivalent to genera in this table.
The genus of the subgenera *Cephalophus* and *Sylvicapra* is *Cephalophus*.
The genus of the subgenera *Neotragus* and *Nesotragus* is *Neotragus*.
The genus of the subgenera *Madoqua* and *Rhynchotragus* is *Madoqua*.
The genus of the subgenera *Raphicerus* and *Nototragus* is *Raphicerus*.
The genus of the subgenera *Gazella, Nanger,* and *Trachelocele* is *Gazella*.
The genus of the subgenera *Kobus, Adenota, Hydrotragus,* and *Onotragus* is *Kobus*.
The genus of the subgenera *Hippotragus* and *Ozanna* is *Hippotragus*.
The genus of the subgenera *Anoa, Bubalus,* and *Syncerus* is *Bubalus*.
The genus of the subgenera *Bibos, Novibos, Bos,* and *Poephagus* is *Bos*.
*** Wild form extinct, and, thus, its behavior is unknown.

Signs and abbreviations:

terr.	= territorial
♂/♀	= male and female in separate territories
♂+♀	= pair in a territory
♂(+♀♀)	= male territorial, (non-territorial) females temporarily in territory (including the special case of a more stable harem in some species)
+	= present
?	= probably present but not beyond any doubt
–	= lacking
(–)	= lacking because species is not territorial (terr.: –)
empty	= unknown at present

In cases where the authors of this book did not know the territorial behavior of the animals under discussion well enough from their own watching, predominantly information from the following authors has been used: Bigalke (1972), Estes (1974), Geist (1971), Gray (1973), Jarman (1974), Jungius (1970), Krämer (1969), Lent (1969), Leuthold (1977), McHugh (1958), Owen-Smith (1977), and Schaller (1977).

Among the territorial bovid species, at least two (possibly three) basically different types of territoriality can be distinguished. Estes (1974) coined the terms "territorial/gregarious" and "territorial/solitary" types (Table 10). Thus, territoriality can be found in otherwise gregarious species as well as in species with solitary life habits, and, depending on whether a species is more gregarious or more solitary, its territorial behavior looks somewhat different.

In the territorial/gregarious type, only adult males become territorial, and there are bachelor groups, female groups, and mixed herds in these species. Although harem behavior may occasionally be combined with territorial behavior, the female groups and the territorial males are separate social units in the majority of cases, and females only temporarily visit males in their territories. Reproduction is more or less restricted to the territorial males and is a major biological function of territoriality in these animals. A more or less regular alternation between territorial and non-territorial periods in the life of the territory owners seems to be typical. Clearly, our Antilopinae species belong to this territorial/gregarious type, and so do the Alcelaphinae and Reduncinae species, and probably also the South African oryx *(Oryx gazella*–Kok pers. comm.). This type also can be expected in some of the bovid species whose territorial behavior has not been investigated to date, and we should perhaps mention that this type of territoriality is not restricted to the Bovidae but occurs as well in such species as pronghorn (*Antilocapra americana*–Bromley 1969).

Essentially, all the differences in territoriality, such as territory size, territorial period length, marking behavior, frequency of interactions among neighbors, etc., which have been pointed out within the Antilopinae species, also occur within the other territorial/gregarious bovid species. Consequently, the similarities in certain aspects of territorial behavior sometimes are much greater among taxonomically relatively distant species than between closer related species. For example, with respect to the relatively small territory size and the nearness of neighboring territories resulting in a dense territorial mosaic, Thomson's gazelle has much more in common with wildebeest, a species from a completely different subfamily of bovids, than with some of the other Antilopinae species such as gerenuk or mountain gazelle. The territories of Coke's hartebeest average roughly 30 ha (Gosling 1974). This size corresponds approximately to that of the territories in Grant's gazelle, but it is considerably more than in white-bearded wildebeest where the territories hardly reach the size of a tommy territory, although wildebeest and hartebeest are much more closely related to each other (both belong to the subfamily Alcelaphinae) than to Grant's gazelle.

On the other hand, there are a few group-specific features and some features recognizable within the Antilopinae, which become more evident when the comparison is broadened to include other territorial/gregarious bovids. For example, none of the territorial waterbuck species (Reduncinae) investigated to date has been found to mark their territories with the secretion of preorbital glands or by establishing dung piles. In the other big groups of territorial/gregarious bovids, the wildebeests, hartebeests and their relatives (Alcelaphinae), at least some of them mark with preorbital gland secretion and all of them establish dung piles. However, urine apparently does not play a role as a means of marking, and the urination-defecation sequence, so typical of adult Antilopinae males, is not found in the Alcelaphinae, Reduncinae, Hippotraginae, etc. To show how sporadically such behaviors show up in taxonomically very distant species, it may be mentioned that precisely the same urination-defecation sequence with the same postures and initiated by pawing the ground with a foreleg as in Antilopinae also occurs in territorial pronghorn males. To give another example, the curious habit of territorial blackbuck males to rest right on top of their dung piles, sporadically occurs in territorial blesbok males (Lynch 1974), but not in other *Damaliscus* species in the Alcelaphinae subfamily.

As in Antilopinae, the combination of territorial behavior with harem behavior appears to be (a) restricted to territorial/gregarious species with relatively large territories, and (b) it only occurs occasionally and is obviously linked to special environmental conditions and/or (small) population density in a given locality. For example, Backhaus (1959) observed a combination of territorial and harem behavior in the comparatively sparse and widely distributed hartebeest population in the Garamba National Park, whereas Gosling (1974) found the female herds crossing through the territories of the males in the much denser hartebeest population of the Nairobi National Park.

In white-bearded wildebeest, adult males stay for months and years in their territories in the Ngorongoro Crater (Estes 1966, 1969). However, in the adjacent Serengeti plains, the territorial behavior of the same subspecies is more or less restricted to the rutting season and the territorial periods of the individual bulls last only for hours or, at maximum, for a few days. In defassa waterbuck, Spinage (1974) found a mean territory size of 202 ha in the Oysa area of the Ruwenzori National Park. On a neighboring peninsula, the mean territory size was only 81 ha, although, as Spinage emphasizes, the vegetation in the two areas appeared to be virtually identical. He relates the greater density of smaller territories on the peninsula to the easier availability of water to the territorial males of this very water-dependent species. In Uganda kob, Buechner (1961) described extremely small territories of 30 m and less in diameter clustered together in "arenas." However, Leuthold (1966, 1977) found considerably larger territories at the periphery of such arenas in the same species and in the same area. Such and similar observations underline the existence of intraspecific variations in territoriality, as we discussed here for the territorial behavior of Grant's gazelle and blackbuck.

Likewise, the principal points in the interactions among territorial neighbors of other territorial/gregarious species appear to be largely the same as those which govern the encounters of territorial bucks in the Antilopinae. Of course, there are very interesting differences in the displays and fighting techniques. However, their consideration belongs more to a comparative discussion of agonistic behavior than of territorial behavior.

Something may perhaps be said here about the role of grazing in agonistic encounters between territorial neighbors. This behavior is present in many, if not all, of the territorial/gregarious bovids. However, with respect to its ritualization into a mutual grazing ritual with a predictable sequence in the changes of the positions of the combatants relative to each other, Thomson's gazelle provides the clearest example presently known. The agonistic grazing in encounters between territorial hartebeest (Gosling 1974) and topi bulls may come nearest to it. In white-bearded wildebeest, grazing plays a great role in territorial encounters (Estes 1969), but here the interactions between the neighbors often do not take place on the boundary. Instead, one combatant may, while grazing, approach the other inside the latter's territory. This situation is entirely different from that in tommy, and on the whole, agonistic grazing seems to play a somewhat different role in the encounters among wildebeest bulls, probably because of the considerably less precise determination of the territorial boundaries in wildebeest. Possibly for the same reason, "excitement activities," particularly self-scratching, appear to be much more frequent in territorial encounters of wildebeest, hartebeest, bontebok *(Damaliscus dorcas dorcas),* etc. (Estes 1969, Gosling 1974, David 1973) than among territorial neighbors in gazelles and their relatives.

In spite of such differences in certain behavioral details however, the agonistic interactions among territorial neighbors are highly ritualized in all the bovid species of the territorial/gregarious type. Sometimes they may result in (minor) changes in the position of the boundaries, but they do not aim to evict the neighbor from his ground. Also, the "dear enemy" relationship, with its positive effects on territoriality, is recognizable in all the species, as it is in Antilopinae.

As with Indian blackbuck (p. 65), Uganda kob in the small arena territories (Buechner 1961) as well as some of the territorial defassa waterbuck (Spinage 1974) also leave their territories and pass through neighboring territories as a daily routine; the kob in order to feed, the waterbuck in order to drink.

In defassa waterbuck with a mean territory size of 80 ha in certain areas and of 200 ha in others (Spinage 1974), the territory size is in about the same range as in mountain gazelle. Until they reach senility, apparently all the fully adult waterbuck males (adulthood is reached at an age of about seven years in this species) are territorial, and no fully adult males are found in the bachelor groups (Spinage 1974). This situation resembles that of gerenuk but not of mountain gazelle. Furthermore, territorial waterbuck males may tolerate bachelor groups within their territories, which is neither true for mountain gazelle nor for gerenuk but was found in Grant's gazelle and Indian blackbuck under certain conditions (p. 176).

In short, when one broadens the comparison of territoriality within the Antilopinae species into a comparison with other territorial/gregarious bovid species, the picture becomes richer, more species-specific differences and a few group-specific differences (e.g., the lack of dung piles in Reduncinae territories) show up, and certain aspects (e.g., the variability of territorial behavior within a species) which are recognizable in some of the Antilopinae species reach a greater degree of certainty.

On the other hand, such a broader comparison does not add many principally new aspects to the picture obtained from the study of Antilopinae. Striking is the relatively frequent occurrence of similarities in single features and in certain details of territorial behavior in taxonomically rather distant species. It is hard to give a truly satisfactory explanation for this fact. Recently Jarman (1974), Leuthold (1977), and others have tried to obtain a more subtle classification of territoriality and social organization in territorial/gregarious bovids by establishing a number of types and subtypes of these organizations related to specific ecological conditions. These ecologically oriented types and subtypes seem approximately to fit the usually one or two species which served as models for their conception. If one tries to apply them to other species, even when one sometimes knows the original species in somewhat more detail than the authors did who created these types, one frequently will find that they do not fit extremely well. Each (new) species under discussion usually falls somewhere in between two or more of the types. In this situation, it appears to be fair for us simply to admit that we have not yet reached a true understanding of the distribution of the different features in the territoriality of the territorial/gregarious bovids, and that the trials to establish a classification of them are only of more or less heuristic value.

COMPARISON WITH BOVID SPECIES OF THE TERRITORIAL/ SOLITARY TYPE

When we speak of a territorial/solitary type (Estes 1974) in bovids, the word "solitary" must not be taken too literally. It does not necessarily imply that each individual lives completely alone. This term also includes species which commonly live in pairs and in which occasionally even a few more animals may be seen together. However, groups of more than two to three members are more or less exceptional in these species. If they occur at all, they usually are in the range of four to five and they hardly ever exceed a maximum of 10 animals. In all these exceptional cases, the groups are not true social units but only very temporary congregations.

In contrast to the territorial/gregarious type, the territories of the territorial/solitary bovids are maintenance territories, i.e., it is not possible to attribute one special function (such as reproduction) to them. Furthermore, in territorial/gregarious species, a great section of a population is non-territorial (e.g., all the females) and spends at least as much time outside of the territories as inside. Even in the life of those individuals who become territorial (adult males), there are long periods during which they do not stay in territories (e.g., when migrating), and, thus, territorial and non-territorial life periods more or less regularly alternate with each other. In species of the territorial/solitary type, however, the overwhelming majority of a population is found inside the territorial space, and the time which an individual may spend outside a territory certainly is much shorter than the time of its stay inside a territory. Consequently, regular and far ranging migrations are incompatible with this form of territoriality, and the territorial periods are lengthy. Provided that the ecological conditions are favorable and do not drastically change in a given locality, it may be assumed that such animals will keep their territories for a lifetime. Finally, these territories are always large, particularly when one takes the usually small size of the inhabitants into account.

The territorial/solitary type occurs in small bovids, such as the Neotraginae (or Neotragini–according to other classification systems) and the duikers (Cephalophinae). It is presently known beyond any possible doubt in only three species. In all the others, it is a more or less likely assumption which needs further investigation. Particularly with respect to this future research, it may be advisable to distinguish two subtypes within the territorial/solitary type. In the one subtype, a pair–male and female–inhabit a mutual territory. This is known for certain from Kirk's dikdik *(Madoqua (Rhynchotragus) kirki)* and klipspringer. In the other subtype, male and female live separate, except for breeding, and each of them has its own territory. This seems to be the case in steenbok (Walther 1972b).

The relatively most detailed information is available on the pair territories in Kirk's dikdik (Simonetta 1966, Tinley 1969, Hendrichs and Hendrichs 1971, Walther 1968b, 1972b). Hendrichs and Hendrichs (1971) estimated the lengths of the territorial stay to be between five and 10 years, i.e., the lifespan. Territory sizes ranging from about 2 to 12 ha have been recorded for these little creatures, and up to 30 ha appear to be possible. Landmarks such as creeks *(korongo)*, trails, rocks, "islands" of bush in otherwise short-grass or semi-arid areas, play an important role in the pair-territories such as dikdik. Both male and female mark with their preorbital glands, although the male marks more frequently, and both use urination-defecation sequence after scraping the ground with a foreleg. Usually the female urinates and defecates first, and the male urinates and defecates after her at the same dung pile. The "defense" of the territory, chasing young males as well as strange females, and (mainly air-cushion) fighting with territorial neighbors, is executed by the male. The speculation is not out of place that the female may be bound primarily to the individual male and only secondarily to the territory. In any case, the male is "more territorial" than the female. In accordance with this impression, a male who has lost a female keeps his territory, but a female who has lost her male cannot keep the territory against the pressure of neighboring pairs in a densely populated area (Hendrichs and Hendrichs 1971). Young dikdik are born and grow up in the parental territory. At an age of about seven months, they are driven out of the parental territory by the father. They wander around in the vicinity singly and try to find an unoccupied area for the establishment of a territory of their own.

As can be seen from this example, the territorial/solitary type differs from the territorial/gregarious type in quite a number of features. On the other hand, there are also some features which resemble the territorial/gregarious type. Among the Antilopinae, the territorial behavior of gerenuk comes relatively close to the territorial/solitary situation as described here for dikdik. Since small body size and more solitary life habits are often considered to be indicative of phylogenetic primitiveness in bovids, one could possibly speculate that the territoriality of the territorial/solitary type represents the original form from which the territoriality of the territorial/gregarious species has evolved. On the other hand, an adherent of the opinion that the territoriality of the territorial/gregarious species is a phylogenetically new invention (e.g., an adaptation to certain negative consequences of gregarious life—see p. 52) which has no immediate connection to the territorial/solitary situation, may also have his arguments. Table 10 gives a summarizing review of territorial behavior in the Bovidae.

12

Management Implications

GENERAL IMPORTANCE OF TERRITORIALITY FOR MANAGEMENT

Territoriality of adult males has been observed in all the gazelles and related species investigated to date. Thus, it certainly has to be considered in the management of all these species. Under management aspects, it is particularly important to keep in mind that (a) the territorial individuals are the guarantors of reproduction, (b) the tendency of adult males to become territorial is very strong, (c) the territorial bucks are dominant over, and quite frequently even absolutely intolerant of, other males, (d) the territorial behavior is variable and adaptable to environmental conditions; however, only within (species-specific) limits, (e) besides its function in reproduction, territoriality has some other functions important to the natural life of these species, and (f) the non-territorial males are also important to the populations. The management implications of these points will now be discussed in greater detail.

Although it is theoretically possible that a non-territorial male can breed a female, it actually has been observed only once in all our studies. Thus, it certainly is rare, and the game manager is well-advised not to rely on this remote possibility. He has to be sure that he has a sufficient number of territorial males in the population under his management. Since a combination of territorial behavior with harem behavior has sometimes been observed in some of the Antilopinae species and it appears likely that this combination may also occasionally occur under natural conditions in other species, keeping females permanently within the territory of a male is not disadvantageous in principle. However, it certainly is more natural to a buck when periods of association with females alternate with periods of a solitary stay in his territory. The opposite seems to be more important in that the manager must watch and–if necessary–insure that the females frequent the territories of the males. Furthermore, it takes a male about a week to fully establish a territory, and he only becomes a true factor in reproduction when he stays in his territory for several weeks or months. Thus, severe disturbances, including management activities (!), of the territorial males and the areas occupied by them should be kept to the absolute unavoidable minimum.

Apart from all theoretical considerations whether territoriality is an innate or a learned behavior in the animals under discussion and whether it should be termed an instinct, a drive, a need, or whatever, a strong tendency to become territorial is always present in adult Antilopinae males. The natural variability in territorial behavior also enables these animals to become territorial under rather artificial and even inadequate conditions if they do not have any choice. Thus, it is to be expected that some of the adult males will at least try to become territorial at almost any time and under almost any conditions. Becoming territorial, however, is regularly linked with a considerable increase of dominance or intolerance toward other males. Consequently, the manager has to make sure that unoccupied space with satisfactory environmental conditions is available in the area for the non-territorial males.

The variability and adaptability of territorial behavior is not unlimited, however. Males may try to become territorial under almost any condition, but they cannot fully establish and, above all, they cannot keep their territories for an appreciable time under unfavorable conditions. For example, a male cannot keep his territorial status when his territory is permanently overrun by huge numbers of males and females, as pointed out in previous chapters. Although territory size may vary with environmental conditions and density of territorial individuals within a given area, there is a (species-specific) minimum size. To name an extreme case which is not rare under captive conditions, one cannot expect several males to become territorial when the space available hardly corresponds to the species-specific minimum size of one territory. Furthermore, there is a difference between minimum size and optimum size. Under natural or semi-natural conditions, the full, positive effect of territoriality on reproduction can only be expected when the territories approach the optimum size.

Although, under management considerations, reproduction is the most important function of territoriality, other aspects are not without practical implications either. The territorial males are the "brakes of migration" (see p. 51) in migratory species. After the herds have arrived in areas with favorable conditions, the continuance of the migration may depend largely on the proportion of territorial males relative to the proportion of non-territorial adult males in the herds (i.e., the "motors of migration," p. 51). Sometimes this aspect may be quite important to a national park with migratory populations which cross its border. Also, the forcing of the bachelor groups into peripheral areas (p. 190) may have management implications. To fulfill these and similar functions, the number of territorial individuals probably has to be considerably larger than the mere fulfillment of reproduction would require.

As important as the territorial males are, non-territorial adult males are not unimportant to the population. In addition to forming the social environment to which males return after a territorial period, they form the reservoir for recruitment of territorial males. Furthermore, the adults in bachelor groups –at least in blackbuck but probably in other Antilopinae species as well–have an amelioratory effect conducive to the social development of young bachelors

(Mungall 1978b). These adults shield the youngsters not only by limiting harassment from other bachelors but also because the adults attract the brunt of aggressive contacts with territorial bucks. Another important social function of non-territorial adult males is as "motors of migration." Finally, non-territorial males apparently pay a considerable toll to natural predation due to their stay in peripheral areas. Thus, they may contribute to a diminution of the predation rate in other sex and social classes including the territorial bucks.

RECOMMENDATIONS FOR NATIONAL PARKS

Provided that the general life requirements (space, food, water, etc.) of the animals under discussion are fulfilled, it follows from the above that the major recommendation for the management of gazelle species or blackbuck in national parks is to give the males a chance for undisturbed territorial behavior. Practically speaking, this means to keep hands off and manage them as little as possible. Particularly in parks with great tourist traffic (or much other traffic), the reproductive success of the territorial males may be diminished when traffic becomes too heavy in the areas occupied by the males. Due to their small flight distances from cars and/or humans, the territorial bucks may establish and keep territories in such areas. However, the females have much larger flight distances (p. 153), and may visit these territories infrequently or even not at all. When they visit these territories, they often will be scared away by the human traffic before the males have a chance to court and breed them. For example, although unfortunately no hard numbers are available, people who have known the Ngorongoro Crater since about 1960 generally have the impression that the population of Thomson's gazelle has continuously declined in the Crater during the past 20 years. This decline coincides with the striking increase of tourist traffic in this area. A very basic and relatively simple management tactic would be to restrict all the traffic, tourist traffic as well as traffic of the staff from workshops, lodges, etc., to the roads and not allow them to cross the country at liberty as practiced there to date. This done, one may possibly have to insure that the road network does not become too dense in the areas with territories.

Provided that undue disturbance is avoided (see above), a national park can take advantage of the territorial behavior of the species under discussion in at least two ways. Most tourists are unaware of territorial behavior in these animals. They often are also not particularly interested in these animals themselves but they look for more spectacular beasts such as lions, elephants, rhinos, etc. However, visitors regularly become interested when the territorial behavior of gazelles is demonstrated to them and when they can observe it themselves which is by no means difficult. Thus, a careful (!) demonstration of territoriality of Antilopinae species to tourists can contribute to raising their interest in the national parks and creating an additional attraction. Secondly, since territorial bucks usually are fine specimens in good physical condition and are easier to approach than any other animals of the same species due to their small flight distances, they are the ideal subjects for photography which certainly is the most frequent and greatest hobby of tourists visiting national parks.

RECOMMENDATIONS FOR HUNTING AREAS

Although it is unlikely that territorial Antilopinae males are proportionally more vulnerable to natural predation than other social classes (p. 153), they certainly are more vulnerable than any other class to human hunting because (a) they are good trophy animals, (b) they can reliably be found at definite places, and (c) they have smaller flight distances than the other conspecifics from humans and vehicles. Thus, the danger that too many territorial males may be taken is always present in any form of modern hunting. This can easily have negative effects on quantity and/or quality of reproduction in the species. In this regard, one also has to keep in mind that a territory which has become vacant, is often not immediately occupied by another male in these species (p. 57).

Although I (Walther) worked exclusively in national parks, I had two very interesting and instructive experiences with hunting problems in Africa. In one case, some professional hunters expressed their disappointment that the population of Thomson's gazelle had continuously declined over the years in their hunting areas, even though only sports hunting was going on there, the natural predation was low, diseases were unknown, and only trophy animals (i.e., males, but no females) were shot. As it turned out in our discussion, pressure from sports hunting was quite high in these areas, and the professional hunters in charge, not to speak of their hunting clients, were unaware of territorial behavior in gazelles. There was hardly any doubt that they shot almost any male as soon as he had established a territory, and that they effectively diminished the reproduction in this way.

In the second case, a formerly protected area was scheduled to be opened to careful hunting. The authorities planned to allow an annual shooting quota of 10% of the gazelle population, and they asked me whether this would be too much. I answered that, provided the size of the population was correctly known, 10% certainly was not too much, but I asked them how the shooting would be distributed among the different sex, age, and social classes of the population. They readily assured me that only trophy animals would be taken, and they were very surprised when I now had severe reservations and objections. Of course, 10% of the population is a small shooting quota when equally distributed over the sex, age, and social classes. However, when the shooting is restricted to trophy animals, only adult males are killed, and they comprise only about a quarter of a gazelle population. Thus, 10% of the population means about 40% of the adult males which is not a small ratio. Since territorial males remain at definite and known locations and have small flight distances, the probability is great that most, if not all, of the adult males killed will be territorial individuals. The proportion of territorial males within a population in a given area may vary with circumstances but averages about 10% (p. 47). Thus, when 10% of the population are taken but only trophy animals are shot, it is very possible and even likely that all the territorial bucks in the area will be killed–with all the ensuing negative consequences on reproduction.

Generally, it seems to be advisable that human hunting should more or less imitate natural predation–of course, with the exception of the killing of fawns –because these animals obviously are adapted to this type of mortality. According to the kills of predators, more adult and subadult males than females should be killed but females should not be completely exempted from hunting. Probably, a shooting quota of ⅔ males and ⅓ females is about right. Solitary females should be protected because in most cases they are mothers of neonates. With respect to their role as a reservoir for territorial males and other social and biological functions, all the adult bachelors should not be taken, of course. Territorial bucks should be completely protected or, at least, as much as possible. This protection simply can be accomplished by not allowing a hunter to shoot any adult buck who is alone or the only adult male with a female group. Sometimes a bachelor may be mistaken for a territorial male and may be saved in this way, and, conversely, the killing of a territorial male is not absolutely prevented by this measure. However, this regulation will work in the majority of cases, and it certainly will diminish the losses of territorial males to a tolerable minimum.

RECOMMENDATIONS FOR GAME RANCHES

On game ranches of the Texas style, in principle the recommendations given for national parks are valid in cases where the animals are displayed to visitors, and the recommendations for hunting areas are valid on ranches where the animals are hunted. However, on the game ranches, additional aspects come into effect which are linked in part to the smaller population sizes, in part to the spatial limitations, and in part to the fact that these animals are not endemic there but have to be introduced with the goal of building up a sizeable population.

In addition to the general aspects of hunting discussed above, intended as well as unintended selection by human hunting plays a greater role on game ranches than in the wild. In the comparatively small populations on spatially limited pastures, it should not be much of a problem for the owner or manager of a ranch to know individually most of the males who are territorial at a given time. If a particular territorial male shows any undesired horn form or other abnormality likely to be hereditary, he should be shot as soon as possible in order to prevent this trait from spreading in the population. On the other hand, a territorial male who shows highly desirable qualities should be protected completely to give him the best chance for transmitting his good qualities to as many descendents as possible. Of course, all the features which make territorial Antilopinae males particularly vulnerable to human (sports) hunting are as valid for hunting on game ranches as they are for hunting in the wild. In the game ranch situation, they may sometimes have even more undesired consequences than in the wild. It is even possible that such negative consequences play a role with respect to the coloration of male blackbuck in Texas since many ranchers complain that only a few of their males show the desired, pitchblack color expected of adult males in this species.

Apparently, several factors play a role in developing this black color, and not all of them are fully understood at present. Provided the subspecies of one's stock represents bucks who can turn black, dark coat color has to do with age since only adult bucks develop a black coat. Also, a connection with season is obvious in many cases, i.e., many males have a black coat in winter but lighten in summer. In some cases, the male's summer coat is only darker on the head than the orange-tan coat of the females and immature males. On the other hand, there are males who remain black all year long. Furthermore, the coloration of the adult bucks also seems–by means of hormonal interaction (Mungall 1978a)–to have something to do with dominance in that a deep black color is most frequently found in very dominant bucks and, thus, particularly often in territorial males. Finally, there seem to be individual differences.

One hunting regulation resulting from these considerations might be to restrict the hunting of blackbuck males to the winter months since more males show the desired black color during this season. Then more dark bachelors would be available to ease pressure on the territorial bucks. However, an alternative approach would be for the manager of a population to exempt all territorial males of maximum black color from hunting for several years in order to favor their genes in the population. Obviously, the opposite "politics" have been practiced on many Texas game ranches. With the desire of a sport hunter for a blackbuck that is black, the availability of such individuals can quickly decrease, causing selection against black color.

Blackbuck are gregarious and a buck without territorial neighbors tends to wander. To keep from disrupting a good breeding area, an alternative hunting site should be sought if fewer than four bucks hold territories in an opening. Furthermore, if too many territorial bucks are removed from one opening in a short time, even an area which has had a continuous series of territories for years will be abandoned in the ranch situation, and new bucks may not take over for several months even after hunting is stopped. Bucks with contiguous territories should not be taken during any one hunt. For each 3 km (about the one-way distance of a female herd's daily circuit) measured along the maximum straight dimension of an opening in which territories are concentrated, one territory should be permanently protected from human hunting. Since a buck tends to return to the same location when he resumes territorial status and since another buck may hold this same ground part of the year, one or two bucks for each 3 km will have a better chance to breed during all the years of their prime if not shot while away from the territory. This procedure should afford the breeding community greater continuity.

In short, as with hunting in the wild, the best procedure is to protect the territorial males of good quality completely. If the ranchers feel that they cannot do so for some reason, the shooting of such males should, at least, be restricted to a minimum until the quality of the population has improved to a desired extent.

Problems of Antilopinae territoriality linked to small pasture size, to feeding from feeders, and to keeping the animals together with other species, are the same as in zoological gardens, and most of them will be discussed in this context (p. 212). However, a word may be said on the introduction of such animals onto rangeland. Our experience in this regard is mainly with blackbuck on Texas ranches; however, *mutatis mutandis,* the principal points certainly are as valid for other species in other areas.

For the full complement of behavior patterns for all social categories to be expressed, the blackbuck population on a ranch needs close to 40 ha. Although they concentrate their activity in flat to rolling grassland, scattered or elongate brush areas are important. If fenced, a single buck can dominate such an area if there is only one opening. In pastures closer to 200 ha in size, this is impossible. Large areas with interconnecting openings allow a large number of territories. Because territorial males without neighbors tend to wander, openings large enough for at least three territories are preferable. Average ranch territories are 4 ha in size and range from 1 ha to 20 ha. In addition, there should be room for bachelor males in the grassy openings. Typically, the bachelors spend at least part of their time within sight of the territories. This facilitates replacement of bucks who are no longer defending their boundaries tenaciously.

Trios consisting of two females and one male are frequently sold for stocking. Starting a new population in an area of 35 ha or more, a manager often releases more animals. Eight blackbuck is the minimum if one wants to favor quick establishment of a normal social order to make success more likely. A favorable distribution of age and sex is three females including at least one fully adult individual, three mature males, and two immature males of one year or somewhat older. The females are the nucleus for a breeding herd. With three females rather than two, an inflexible superior-inferior relationship that might interfere with reproductive success is less likely to form. The three adult males furnish at least one territorial buck. If more than one become territorial, then each territorial buck will probably have one or two territorial neighbors. If not all become territorial, then the bachelor association formed by the two immature males will profit by adult leadership. By the time the immature males grow old enough to hold territories themselves, the possibility of there being a male fawn with no male age-mates and no bachelor association to join when forced away from the females will have decreased considerably.

Some ranches support their game populations above carrying capacity by supplemental feeding. Blackbuck pour into the feeding area, but only adult bucks push through the mass of other species initially at the feeder. Males spend part of their time displaying to the assembled females instead of trying to eat. By the time a territorial male in the feeder area abandons displaying to the other blackbuck, little or no feed is left. To relieve crowding and to make feed available to more animals, the feed should be spread out rather than placed in one trough. Then, even blackbuck females can dart in and get something to eat. If a territorial buck is preventing bachelors from using a feeding site within his territory, then another feeding site should be established on unoccupied ground.

RECOMMENDATIONS FOR ZOOLOGICAL GARDENS

As mentioned above, the conditions as well as the recommendations for keeping Antilopinae on rangeland in small pastures approximate the conditions and recommendations for keeping these animals in zoological gardens. Usually, the zoo pens are only a fraction of 1 ha in size. Zoo pens of 2 to 3 ha or even only 1 ha are extremely rare, and one almost never sees zoo enclosures which exceed a maximum of 3 ha. Thus, even in species such as blackbuck and Thomson's gazelle, where territory size is comparatively small, only one male can become territorial in such an enclosure. The probability that he will become territorial is extremely great because the tendency to establish a territory is strong and always present in these animals, it is flexible enough to adapt to captive conditions, and it is favored by the zoo situation in that migrations are excluded, sufficient food and water are readily available, and usually some females are always with the male. Under these conditions, the male takes all the space available–the outdoors enclosure as well as the stall–as his territory, and even in species with comparatively short territorial periods under natural conditions, he usually remains territorial all the time. If there are conspecific males above the age of adolescence, but sometimes even below it, he harasses them more or less continually. Besides fighting and chasing them, which can be much more detrimental under the spatial limitations of captivity than in the wild, he may prevent them from feeding, resting, and using shelter from cold, heat, rain, etc. In short, a male conspecific kept together with a territorial buck under the usual zoo conditions is condemned to die sooner or later if left in the enclosure permanently. One can take certain temporary and limited measures which make life easier for the non-territorial males and which may somewhat diminish the probability of a fatal ending. However, when the enclosure is smaller than a minimum of 1 ha, the measures will not work in the long run. The most important of these measures are: (a) avoidance of all sharp and right angle corners in the fence so that a subordinate male cannot become cornered when withdrawing or fleeing from the territorial male, (b) arrangement of several widely distributed feeding places in the enclosure which should be filled with food simultaneously so that the subordinates have a chance to feed relatively undisturbed, (c) establishment of some structure in the enclosure by trees, rocks, ditches, etc., which provides some cover for the subordinates so that they can disappear, at least temporarily, from the eyesight of the territorial buck, and (d) separation of the males in separate boxes when brought into a stall.

As stated above, one can make the situation more tolerable for the non-territorial males even in a relatively small enclosure in this way for a short time. Chances that these measures will work for a longer time increase when they are applied in pens larger than 1 ha. However, a guarantee that the keeping of one or several non-territorial males together with a territorial buck will not have any detrimental effect on the bachelors is only given when the enclosure is essentially larger than the size of a territory, and this is practically impossible in a zoo.

Provided that the enclosure is at least 1 ha in size, another measure to diminish the negative effects of the one male's territoriality upon the others can be tried. However, it must always be used in combination with all the measures listed above. A group of gazelle or blackbuck can be kept together with animals of another, bigger, non-territorial, but otherwise rather active, ungulate species. Under these conditions, of course, one will have to face all the problems of keeping different species together (see below); however, the negative effects of intraspecific territoriality can be considerably diminished in this way. When a group of Indian blackbuck was kept together with wild goats (markhor and ibex) in a large enclosure with trees, rocks, many well-distributed feeders, etc., the blackbuck still got their food and had sufficient shelter (Walther 1959). On the other hand, the wild goats somewhat dominated and disturbed them so that they were permanently on the move and none of the males became fully territorial. Since the females were permanently together with the males due to the captive situation, the males succeeded in breeding them. In addition to females, there were two fully adult bucks and several immature males in this blackbuck group. They were kept together with the wild goats for two years, and there were no cases of mortality or severe damage due to intraspecific aggression during this time. Later, when the wild goats were separated from the blackbuck, one of the two adult bucks immediately became territorial and gave the other males such a hard time that they had to be removed from the enclosure. Thus, it is proven that this method works. Due to the problems which arise from keeping different ungulate species together in captivity, however, it somewhat resembles the method of driving out a devil by another devil.

To date, no zoological garden has tried to imitate the natural situation of an all-male group, i.e., to keep a whole group of conspecific and more or less equal males without any females in a large and well-structured enclosure. Because of lack of practical experience, we cannot say whether one of these males will eventually become territorial and will harass the others under these conditions; it is absolutely possible. On the other hand, it is just as possible that, for a long while, the bucks might consume their aggressive potentials continuously by numerous but harmless sparring among each other, and that none of them might become territorial and, thus, truly dangerous to the others. Of course, there would be no reproduction in such an all-male group, but a whole group of fine males with their interesting activities could be a beautiful display for a zoo. As stated above, the outcome is unknown and, theoretically, the chances are fifty-fifty in the best case. However, some chance exists, and it would be quite worthwhile to conduct such an experiment.

Under captive conditions, some problems occasionally arise with respect to females and young. When no conspecific males or substitutes (see below) are available, a territorial male may eventually direct his unsatisfied aggressive tendencies toward females or young. It seems that species in which the males have a largely different behavioral inventory toward females than among each other (p. 136), are less inclined to do so than species in which the males' behavior toward the females is more or less the same as the behavior of males toward each other. Generally, such events are infrequent in all the species under discussion. Nevertheless, an adult buck should always be separated from the females when in a stall. Occasionally under captive conditions, a male may intensively court a female during the days before she gives birth, i.e., he can bother her quite considerably at a time when she would be better undisturbed. Therefore, it is advisable temporarily to remove either this female or the territorial male from the enclosure.

Also for other reasons, e.g., for timing fawning at a favorable season, it sometimes is advisable to separate the breeding territorial male from the females for a predetermined period each year. As can be taken from the descriptions of the situation in the wild, it is by no means unnatural for an Antilopinae male to be alone in his territory periodically. However, one should separate the male from the females in such a way that he cannot see them all the time. Besides the probability of psychological stress and frustration, the male may become so restless when he can see the females all the time but cannot go to them, that he may lose weight and condition.

The greatest problems arise from the aggressive tendencies of territorial males in captivity. Due to his aggressiveness toward conspecific males, such a buck is commonly kept only together with his females in most zoological gardens. As stated above, aggression toward females is exceptional in these animals. Under natural conditions, they are relatively most frequent during herding actions, but herding of females usually does not take place in the pen situation. On the other hand, territoriality is linked with a strong increase of aggressiveness which is needed for encounters with neighbors and for driving away trespassing bachelors in the wild. Under captive conditions, the male has no outlet for his aggression, it becomes dammed up, and he starts to attack substitutes. These substitutes may be inanimate objects, animals of another species, or humans.

The most tolerable form, of course, is the aggression toward inanimate objects. As we know from the previous chapters, object aggression belongs to the natural behavior of these animals. Thus, it is not a new behavior; only its frequency may increase under captive conditions. Provided that there are no objects which are fragile or with which the male can injure himself, not much harm can be done by object aggression. If the material of the object is very hard, as with iron fence posts or cement walls, the male may wear down his horns in the course of time. However, the horns remain in good shape when one offers objects of a softer material such as wooden branches. The male will readily take to them, particularly when one renews them from time to time and/or changes their location within the pen. This object aggression has very positive aspects. In fighting the inanimate objects, the male can get rid of at least part of his aggressive tendencies which he otherwise would direct toward living beings. Therefore, several tree branches simply belong to the standard equipment for any pen in which such a male is kept. Many people who keep such animals, complain about their aggressiveness. Apparently, these people do not know how easily and cheaply they can provide an outlet for this aggression and at least somewhat diminish the frequency of aggression toward other animals and humans in this way.

Aggressive interactions with other species, of course, only take place when the latter are kept together with the Antilopinae in the same or in an adjacent enclosure. It must be emphasized that there are more problems in keeping ungulates of different species together than arise from territoriality and that the Antilopinae usually are the ones which suffer in these interactions because usually the animals of the other species are bigger and stronger. Also, the Antilopinae may not always be the "troublemakers;" the other species may harass them without being provoked. There is interspecific competition over food, resting places, shelter, etc., in captivity. Some species such as zebra also can become dangerous to the antelope calves and fawns under captive conditions (Walther 1964c). However, these various difficulties of keeping different ungulate species together will not be discussed in detail in this book. Only the problems which arise from the territorial behavior of the Antilopinae species are of interest here.

Hediger (1942, 1954) spoke of an "assimilation tendency" or "zoomorphism" in animals. Zoomorphism means that under certain conditions, an animal may treat an animal of a different species as if it were a conspecific. Individual familiarity certainly is a factor which enormously favors this assimilation tendency. Thus, although it is not absolutely lacking in the wild (Walther 1979), the assimilation tendency is particularly frequent and pronounced under captive conditions when animals of different species are kept together constantly and in spatial limitations which simply force them to become familiar with each other. In relatively rare cases the assimilation tendency of captive animals results in friendly relationships such as mutual grooming, somewhat more frequently in sexual interactions, but most frequently in agonistic encounters. Apparently, the animals can usually recognize the sex of the individuals of the other species, because such agonistic encounters are primarily between the males of the different ungulate species. In short, species which do not, or which only very exceptionally, interact with each other in the wild may frequently fight each other when kept together in captivity. At least among ungulates, differences in body size, horn size and shape, color patterns of the coat, etc., do not play any role and aggressive interactions between such unequal combatants as dorcas gazelle and eland antelope, Thomson's gazelle and giraffe, blackbuck and axis deer (Figure 73), etc., have been observed, sometimes with disasterous consequences to the weaker fighter. However, it is not always the larger animal which challenges the smaller opponent and starts the fight. The opposite is at least as frequent, particularly with territorial gazelle males. If such a buck is kept together with other bovids, cervids, etc., in an enclosure which he considers to be his territory, he may display to the males of the other species and fight and chase them as if they were intruders of his own species.

Figure 73: Interspecific interactions in captive animals. Blackbuck fighting axis stag over a feeder on a Texas ranch. (Photo: E.C. Mungall—Quajolote Ranch, Texas.)

The frequency and severity of such interactions depend on a variety of circumstances. However, the possibility is always present, and frequent and severe interactions of this type are by no means rare. Certainly, the animal keeper is well-advised when he is prepared and takes precautions for the worst situation.

Furthermore, this assimilation tendency can also be extended to humans. Particularly when a male has been raised by bottle and has become "imprinted" to people, and even when he has become sufficiently familiar with people during his later life, he can treat them as conspecifics. Thus, he displays to them and fights them as if they were territorial rivals or were trespassing bachelors of his own species. In some cases, this aggression is restricted to a few definite persons, often those who have become especially familiar to the animal. In other cases, the animal is aggressive toward any human, and the person's sex does not play any role. Such a buck may even try to fight the zoo visitors through the fence. Then, particularly with respect to children, one has to prevent the visitors from coming too close, e.g., by double fencing. Otherwise, the buck is usually more endangered than the visitors in that he may break his horns when goring and twisting through the fence. Attacks of territorial Antilopinae males toward members of the zoo staff who have to enter the enclosures usually are not truly dangerous due to the small size of these animals, but they are bothersome and can easily result in damaged clothes. The best way to avoid such aggression toward humans is to prevent the males from becoming too familiar with people. Particularly, one should avoid raising a young male by bottle. If it cannot be avoided, one has to be prepared for this male to become aggressive toward people when grown up and especially when he has become territorial–and not least of all against those persons who once gave him the bottle. When such a male is aggressive only toward a few people or only one definite man, it sometimes helps to entrust the male's care to another person. However, one must never forget that these animals are individuals and react as individuals, and they sometimes establish individual attachments or antipathies toward definite persons. Therefore, no general rule is without exceptions. In a reported case (Walther 1966), for example, a captive dorcas buck would attack any known or strange man who entered his pen, except one definite person who could do almost anything with him.

Bibliography

Aeschlimann, A. 1963. Observations sur *Philantomba maxwelli* (Hamilton-Smith) une Antilope de la forêt éburnée. *Acta Tropica* 20:341-368.

Allen, G. 1939. A checklist of African mammals. *Bull. Mus. Comp. Zool.* 83:1-763.

Allison, J.E., Dittmar, G.W. and Hensell, J.L. 1975. Soil survey of Gillespie County, Texas. U.S.D.A., S.C.S. in coop. Texas Agric. Exp. St.

Altmann, D. 1969. *Harnen und Koten bei Säugetieren.* Wittenberg-Lutherstadt; A. Ziemsen.

Ansell, W.F.H. 1971. Proboscidea, Perissodactyla, Artiodactyla. *In: The mammals of Africa–an identification manual,* parts 11, 14, and 15, ed. J. Meester and H.W. Setzer. Washington: Smithsonian Inst.

Antonius, O. 1939. Über Symbolhandlungen und Verwandtes bei Säugetieren. *Z. Tierpsychol.* 3:263-278.

Backhaus, D. 1958. Beitrag zur Ethologie der Paarung einiger Antilopen. *Z. Zuchthygiene.* 2:281-293.

Backhaus, D. 1959. Beobachtungen über das Freileben von Lelwel-Kuhantilopen *(Alcelaphus buselaphus lelwel,* Heuglin 1877) und Gelegenheitsbeobachtungen an Sennar-Pferdeantilopen (*Hippotragus equinus bakeri,* Heuglin 1863). *Z. Säugetierk.* 24:1-34.

Backhaus, D. 1961. *Beobachtungen an Giraffen in Zoologischen Gärten und in freier Wildbahn.* Bruxelles: Inst. Parcs Nat. Congo.

Baharav, D. 1974. Notes on the population structure and biomass of the mountain gazelle, *Gazella gazella gazella. Israel J. Zool.* 23:39-44.

Baldwin, J.H. 1876. *Large and small game of Bengal and the north-western provinces of India.* London: Henry S. King & Co.

Benz, M. 1973. Zum Sozialverhalten der Sasin (Hirschziegenantilope, *Antilope cervicapra* L. 1758). *Zool. Beiträge* 19:403-466.

Bigalke, R.C. 1970. Observations on springbok populations. *Zoologica Africana* 5:59-70.

Bigalke, R.C. 1972. Observations on the behavior and feeding habits of the springbok, *Antidorcas marsupialis. Zoologica Africana* 7:333-359.

Blanford, W.T. 1888-91. *The fauna of British India, including Ceylon and Burma, Mammalia.* London: Taylor & Francis.

Boetticher, H. v. 1953. Gedanken über eine natürliche systematische Gruppierung der Gazellen. *Z. Säugetierk.* 17:83-92.

Brander, A.A.D. 1923. *Wild animals in central India.* London: Edward Arnold & Co.

Bromley, P.T. 1969. Territoriality in pronghorn bucks on the National Bison Range, Boise, Montana. *J. Mammal.* 50:81-89.

Brooks, A. 1961. *A study of Thomson's gazelle (Gazella thomsoni Günther) in Tanganyika.* London: H. M. Stat. Office.

Buechner, H.K. 1961. Territorial behavior in Uganda kob. *Science* 133:698-699.

Burt, W.H. 1943. Territoriality and home range concepts as applied to mammals. *J. Mammal.* 24:346-352.

Carr, J.T. 1969. *The climate and physiography of Texas,* rep. 53. Texas Water Dev. Board.

Cary, E.R. 1976a. Blackbuck menu. *Texas Pk. & Wildl.* 34:16-18.

Cary, E.R. 1976b. Territorial and reproductive behavior of the blackbuck antelope *(Antilope cervicapra).* Unpubl. Ph.D. dissertation. Texas A&M Univ., College Station, Tex.

Crandall, L.S. 1964. *The management of wild animals in captivity.* Chicago: Univ. Chicago Press.

Cronwright-Schreiner, S.C. 1925. *The migratory springbucks of South Africa.* London: T. Fisher.

David, J.H.M. 1973. The behaviour of the bontebok, *Damaliscus dorcas dorcas* (Pallas 1766), with special reference to territorial behaviour. *Z. Tierpsychol.* 33:38-107.

Dharmakumarsinhji, K.S. 1967, Browsing behaviour of *Gazella g. bennetti* and *Antilope cervicapra* in captivity and natural habitat. *Int. Union For. Res. Organ., XIV Congr., sect.* 26:424-465.

Dharmakumarsinhji, K.S. 1978. Velavadar National Park, Gujarat, India. *Tigerpaper* 5:6-80.

Dittrich, L. 1965. Absetzen von Voraugendrüsensekret an den Hörnern von Artgenossen bei Gazellen und Dikdiks. *Säugetierkdl. Mitt.* 13:145-146.

Ellermann, J. and Morrison-Scott, T.C.S. 1951. *Checklist of Palaearctic and Indian mammals.* London: Trust. Brit. Mus.

Emlen, J.T. 1957. Defended area? A critique of the territory concept and of conventional thinking. *Ibis* 99:352.

Estes, R.D. 1966. Behaviour and life history of the wildebeest *(Connochaetes taurinus* Burchell). *Nature* 212:999-1000.

Estes, R.D. 1967. The comparative behavior of Grant's and Thomson's gazelles. *J. Mammal.* 48:189-209.

Estes, R.D. 1969. Territorial behavior of the wildebeest *(Connochaetes taurinus* Burchell, 1823). *Z. Tierpsychol.* 26:284-370.

Estes, R.D. 1972. The role of the vomeronasal organ in mammalian reproduction. *Mammalia* 36:315-341.

Estes, R.D. 1974. Social organization of the African Bovidae. *In: The behaviour of ungulates and its relation to management,* ed. V. Geist and F.R. Walther, pp. 166-205. IUCN Publ. No. 24. Morges: IUCN.

Etkin, W. 1964. *Social behavior and organization among vertebrates.* Chicago: Univ. Chicago Press.

Ewer, R.F. 1968. *Ethology of mammals.* New York: Plenum Press.

Fisher, J. 1954. Evolution and bird sociality. *In: Evolution as a process,* ed. J. Huxley. London: Allen & Unwin.

Geist, V. 1971. *Mountain sheep: a study in behavior and evolution.* Chicago: Univ. Chicago Press.

Gentry, A.W. 1964. Skull characters of African gazelles. *Ann. Mag. Nat. Hist.* 7:353-382.

Gentry, A.W. 1971. Genus *Gazella. In: The mammals of Africa–an identification manual,* part 15.1, ed. J. Meester and H.W. Setzer. Washington: Smithsonian Inst.

Gosling, L.M. 1974. The social behaviour of Coke's hartebeest *Alcelaphus buselaphus cokei. In: The behaviour of ungulates and its relation to management,* ed. V. Geist and F.R. Walther, pp. 488-511. IUCN Publ. No. 24. Morges: IUCN.

Grau, G.A. 1974. Behavior of mountain gazelle in Israel. Unpubl. Ph.D. dissertation, Texas A&M Univ., College Station, Texas.

Grau, G.A. and Walther, F.R. 1976. Mountain gazelle agonistic behaviour. *Anim. Behav.* 24:626-636.

Gray, D.R. 1973. Social organization and behavior of muskoxen *(Ovibos moschatus)* on Bathurst Island, N. W. T. Unpubl. Ph.D. dissertation, Univ. Alberta, Edmonton, Canada.

Groves, C.P. 1969. On the smaller gazelles of the genus *Gazella* de Blainville, 1816. *Z. Säugetierk.* 34:38-60.

Groves, C.P. 1972. Blackbuck. *Encyclop. Animal World* 3:224.

Grzimek, B. and Grzimek, M. 1960. A study of the game of the Serengeti Plains. *Z. Säugetierk.* 25, Sonderheft.

Haltenorth, T. 1963. Klassifikation der Säugetiere: Artiodactyla. *Handb. Zool.* 81(18):11-167.

Harmel, D.E. 1975. *Habitat preferences of exotics.* Kerr Wildl. Manage. Area Res., Federal Aid. Proj. No. W-76-R-18, job performance rep. Texas Pk. & Wildl. Dept.

Hediger, H. 1934. Zur Biologie und Psychologie der Flucht bei Tieren. *Biol. Zentralbl.* 54:21-40.

Hediger, H. 1940. Zum Begriff der biologischen Rangordnung. *Rev. Suisse Zool.* 47:3.

Hediger, H. 1941. Biologische Gesetzmässigkeiten im Verhalten von Wirbeltieren. *Mitt. Naturf. Ges. Bern:*37-55.

Hediger, H. 1942. *Wildtiere in Gefangenschaft.* Basel: Schwabe. English ed. 1950. *Wild animals in captivity.* London: Butterworth.

Hediger, H. 1949. Säugetier-Territorien und ihre Markierung. *Bijdr. Dierk.* 28:172-184.

Hediger, H. 1951. *Observations sur la psychologie animale dans les parks nationaux du Congo Belge.* Bruxelles: Inst. Parcs Nat. Congo.

Hediger, H. 1954. *Skizzen zu einer Tierpsychologie im Zoo und im Zirkus.* Stuttgart: Europa. English ed. 1955. *Studies of the psychology and behaviour of captive animals in zoos and circuses.* London: Butterworth.

Hendrichs, H. 1970. Schätzungen der Huftierbiomasse in der Dornbuschsavanne nördlich und westlich der Serengetisteppe in Ostafrika nach einem neuen Verfahren und Bemerkungen zur Biomasse der anderen pflanzenfressenden Tierarten. *Säugetierkdl. Mitt.* 18:237-255.

Hendrichs, H. and Hendrichs, U. 1971. *Dikdik und Elefanten.* München: Piper.

Herlocker, D. 1975. *Woody vegetation of the Serengeti National Park.* College Station: Kleberg Studies in Natural Resources.

Hvidberg-Hansen, H. and De Vos, A. 1971. Reproduction, population and herd structure of two Thomson's gazelle populations *(Gazella thomsoni* Günther). *Mammalia* 35:1-16.

Jarman, P.J. 1974. The social organisation of antelope in relation to their ecology. *Behaviour* 48:215-267.

Jerdon, T.C. 1874. *The mammals of India.* London: John Wheldon.

Johnson, J.M. 1975. The blackbuck in Point Calimere Sanctuary, Tamil Nadu, population dynamics and observations on behaviour. *Indian Forester* 101:484-494.

Jungius, H. 1970. Studies on the breeding biology of the reedbuck *(Redunca arundinum* Boddaert, 1785) in the Kruger National Park. *Z. Säugetierk.* 35:129-146.

Karunakaran, M.S. 1972. *Working plan for the Thanjavur Forest Division, 1st April 1971 to 31st March 1981.* Madras: Office of the Chief Conservator of Forests.

Klingel, H. 1967. Soziale Organisation und Verhalten freilebender Steppenzebras. *Z. Tierpsychol.* 25:580-624.

Knappe, H. 1964. Zur Funktion der Jacobsonschen Organs. *D. Zoolog. Garten* 28:188-194.

Krämer, A. 1969. Soziale Organisation und Sozialverhalten einer Gemspopulation *(Rupicapra rupicapra* L.) der Alpen. *Z. Tierpsychol.* 26:889-964.

Krishnan, M. 1971. An ecological survey of the larger mammals of peninsular India, lst part. *J. Bombay Nat. Hist. Soc.* 68:503-536.

Kruuk, H. 1972. *The spotted hyena.* Chicago: Univ. Chicago Press.
Lange, J. 1972. Studien an Gazellenschädeln. Ein Beitrag zur Systematik der kleineren Gazellen, *Gazella* (De Blainville, 1816). *Säugetierkdl. Mitt.* 20:193-249.
Lent, P.C. 1969. A preliminary study of the Okavango lechwe (*Kobus leche leche* Gray). *E. Afr. Wildl. J.* 7:147-157.
Leuthold, W. 1966. Variations in territorial behaviour of Uganda kob, *Adenota kob thomasi* (Neumann, 1896). *Behaviour* 27:214-258.
Leuthold, W. 1971. Freilandbeobachtungen an Giraffengazellen *(Litocranius walleri)* im Tsavo-Nationalpark, Kenia. *Z. Säugetierk.* 36:19-37.
Leuthold, W. 1973. Notes on the behaviour of two young antelopes reared in captivity. *Z. Tierpsychol.* 32:418-424.
Leuthold, W. 1977. *African ungulates.* Berlin, Heidelberg, New York: Springer.
Leuthold, W. 1978a. On social organization and behaviour of the gerenuk *Litocranius walleri* (Brooke 1878). *Z. Tierpsychol.* 47:194-216.
Leuthold, W. 1978b. On the ecology of the gerenuk *Litocranius walleri* (Brooke 1878). *J. Anim. Ecol.* 47:471-490.
Lind, H. 1959. The activation of an instinct caused by a "transitional action." *Behaviour* 14:123-135.
Lorenz, K. 1963. *Das sogenannte Böse.* Wien: Borotha-Schoeler. English ed. 1966. *On aggression.* New York: Bantam Books, Inc.
Lydckker, R. 1907. *The game animals of India, Burma, Malaya, and Tibet.* London: Rowland Ward.
Lynch, C.D. 1974. A behavioural study of blesbok, *Damaliscus dorcas phillipsi,* with special reference to territoriality. *Mem. Nas. Mus. Bloemfontain* 8:1-83.
Mason, D.R. 1976. Some observations on social organization and behaviour of springbok in the Jack Scott Nature Reserve. *S. Afr. J. Wildl. Res.* 6:33-39.
McDougall, W. 1923. *An outline of psychology.* London: Methuen.
McHugh, T. 1958. Social behavior of the American buffalo *(Bison bison bison). Zoologica* 43:1-40.
Meester, J. 1959. Some notes on the dibatag. *Afric. Wildl.* 13:281-283.
Mendelssohn, H. 1972. On the biology and ecology of gazelles in Israel. Unpubl. final report on work carried out under the Smithsonian grants Nos. SFG-1-7066 and SFG-0-5181.
Mendelssohn, H. 1974. The development of the populations of gazelles in Israel and their behavioural adaptations. *In: The behaviour of ungulates and its relation to management,* ed. V. Geist and F.R. Walther, pp. 722-743. IUCN Publ. No. 24. Morges: IUCN.
Morrison-Scott, T.C.S. 1939. Some Arabian mammals collected by Mr. H. St. J.B. Philby. *C.I.E., Novit. Zool.* 41:181-211.
Moynihan, M. 1955. Some aspects of reproductive behaviour in the black-headed gull *(Larus ridibundus ridibundus* L.) and related species. *Behaviour,* Suppl. 4.
Mungall, E.C. 1978a. *The Indian blackbuck antelope: a Texas view.* College Station: Kleberg Studies in Natural Resources.
Mungall, E.C. 1978b. Social development of the young blackbuck antelope *(Antilope cervicapra). AAZPA Regional Workshop Proceedings 1977-78*:153-170.
Mungall, E.C. 1979. Effect of space limitations on behavior expressed by blackbuck antelope (*Antilope cervicapra* L. 1758). *Environmental Awareness* 2:41-53.
Mungall, E.C. 1980. Courtship and mating behavior of the dama gazelle (*Gazella dama* Pallas 1766). *D. Zoolog. Garten* 50:1-14.
Natarajan, K.N. and Sunderraj, T.S. 1977. Blackbuck *(Antilope cervicapra)* census in Point Calimere. Unpubl. report of the Wildlife Conservation Society of Tiruchirapalli. 4 pp.
Nice, M.M. 1941. The role of territory in bird life. *Amer. Midl. Nat.* 26:441-487.

Noble, G.K. 1939. The role of dominance in the life of birds. *Auk* 56:263-273.
Owen-Smith, R.N. 1975. The social ethology of the white rhinoceros, *Ceratotherium simum* (Burchell, 1817). *Z. Tierpsychol.* 38:337-384.
Owen-Smith, R.N. 1977. On territoriality in ungulates and an evolutionary model. *The Quart. Rev. Biol.* 52:1-38.
Oza, G.M. 1976. The blackbuck of Baroda. *Sci. Today* 10:49.
Prakash, I. 1975. The amazing life in the Indian desert, our birds, animals, trees. *The Illustr. Weekly of India Annual* 75:96-121.
Prasad, N.L.N.S. 1981. Home range, dispersal and movement of blackbuck *(Antilope cervicapra* L.) population in relation to seasonal changes in Mudmal and environs. Unpubl. Ph.D. thesis. Osmania Univ. Hyderbad (A.P.), India.
Prater, S.H. 1971. *The book of Indian animals.* 3rd (rev.) ed. Bombay: Bombay Nat. Hist. Soc.
Ralls, K. 1969. Scent-marking in Maxwell's duiker, *Cephalophus maxwelli. Americ. Zool.* 9:1071.
Ralls, K. 1974. Scent-marking in captive Maxwell's duiker. *In: The behaviour of ungulates and its relation to management,* ed. V. Geist and F.R. Walther, pp. 114-123. IUCN Publ. No. 24. Morges: IUCN.
Ralls, K. 1975. Agonistic behaviour in Maxwell's duiker, *Cephalophus maxwelli. Mammalia* 39:241-249.
Raychaudhuri, S.P., Agarwal, R.R., Datta Biswas, N.R., Grupta, S.P. and Thomas, P.K. 1963. *Soils of India.* New Delhi: Indian Council of Agricultural Research.
Robinette, W.L. and Archer, A.L. 1971. Notes on aging criteria and reproduction of Thomson's gazelle. *E. Afr. Wildl. J.* 9:83-98.
Schaller, G.B. 1967. *The deer and the tiger.* Chicago: Univ. Chicago Press.
Schaller, G.B. 1972. *The Serengeti lion.* Chicago: Univ. Chicago Press.
Schaller, G.B. 1977. *Mountain monarchs.* Chicago: Univ. Chicago Press.
Schenkel, R. 1966. Zum Problem der Territorialität und des Markierens bei Säugetieren –am Beispiel des Schwarzen Nashorns und des Löwen. *Z. Tierpsychol.* 23:593-626.
Schmied, A. 1973. Beiträge zu einem Aktionssystem der Hirschziegenantilope (*Antilope cervicapra* Linné 1758). *Z. Tierpsychol.* 32:153-198.
Schneider, K.M. 1931. Das Flehmen (2. Teil). *D. Zoolog. Garten* 4:349-364.
Schwarz, E. 1937. Die fossilen Antilopen von Oldoway. *Wiss. Erg. Oldoway-Exped. 1913, N.F.* 4:8-104.
Simonetta, A.M. 1966. Osservazioni etologiche et ecologiche sui dikdik (gen. *Madoqua:* Mammalia, Bovidae) in Somalia. *Mon. Zool. It.,* Suppl.:1-33.
Simpson, G.G. 1945. The principles of classification and a classification of mammals. *Bull. Amer. Mus. Nat. Hist.* 85:1-350.
Sinclair, A.R.E. and Norton-Griffiths, M. 1979. *Serengeti: dynamics of an ecosystem.* Chicago: Univ. Chicago Press.
Sokolov, I. 1953. *Trial of a natural classification of Bovidae.* (Russian). Moscow: Acad. Publ. Moscow.
Spinage, C.A. 1974. Territoriality and population regulation in the Uganda defassa waterbuck. *In: The behaviour of ungulates and its relation to management,* ed. V. Geist and F.R. Walther, pp. 635-643. IUCN Publ. No. 24, Morges: IUCN.
Stockley, C.H. 1928. *Big game shooting in the Indian Empire.* London: Constable & Co., Ltd.
Tinbergen, N. 1953. Social behaviour in animals. London: Butler & Tanner.
Tinley, K.L. 1969. Dikdik *Madoqua kirki* in South West Africa: notes on distribution, ecology, and behaviour. *Madoqua* 1:7-33.
Ullrich, W. 1963. Einige Beobachtungen an Wildtieren in Ostafrika II. *D. Zoolog. Garten* 27:187-193.

Walther, F.R. 1958. Zum Kampf- und Paarungsverhalten einiger Antilopen. *Z. Tierpsychol.* 15:340-380.

Walther, F.R. 1959. Beobachtungen zum Sozialverhalten der Sasin (Hirschziegenantilope, *Antilope cervicapra* L.). *Jahrb. G. v. Opel-Freigehege* 2:64-78.

Walther, F.R. 1960a. Beobachtungen über die Mutter-Kind-Beziehungen und die Rolle des Hengstes bei der Aufzucht eines Fohlens von *Equus quagga boehmi. Jahrb. G. v. Opel-Freigehege* 3:44-52.

Walther, F.R. 1960b. Einige Verhaltensbeobachtungen am Bergwild des Georg von Opel-Freigeheges. *Jahrb. G. v. Opel-Freigehege* 3:53-89.

Walther, F.R. 1961. Zum Kampfverhalten des Gerenuk *(Litocranius walleri). Nat. u. Volk.* 91:313-321.

Walther, F.R. 1963a. Einige Verhaltensbeobachtungen am Dibatag *(Ammodorcas clarkei* Thomas, 1891). *D. Zoolog. Garten* 27:233-261.

Walther, F.R. 1963b. *Juni-Monatsbericht der Georg von Opel-Freigeheges:* 1-6.

Walther, F.R. 1964a. Einige Verhaltensbeobachtungen an Thomsongazellen *(Gazella thomsoni* Günther, 1884) im Ngorongoro-Krater. *Z. Tierpsychol.* 21:871-890.

Walther, F.R. 1964b. Zum Paarungsverhalten der Sömmeringgazelle *(Gazella soemmeringi* Cretzschmar, 1826). *D. Zoolog. Garten.* 29:145-160.

Walther, F.R. 1964c. Ethological aspects of keeping different species of ungulates together in captivity. *Internat. Zoo. Yearbk.* 5:1-13.

Walther, F.R. 1965. Verhaltenstudien an der Grantgazelle (*Gazella granti* Brooke, 1872) im Ngorongoro-Krater. *Z. Tierpsychol.* 22:167-208.

Walther, F.R. 1966. *Mit Horn und Huf.* Berlin, Hamburg: Paul Parey.

Walther, F.R. 1967. Huftierterritorien und ihre Markierung. *In: Die Strassen der Tiere,* ed. H. Hediger, pp. 26-45. Braunschweig: Vieweg.

Walther, F.R. 1968a. *Verhalten der Gazellen.* Wittenberg-Lutherstadt: A Ziemsen.

Walther, F.R. 1968b. Ducker, Böckchen und Waldböcke. *In: Grzimeks Tierleben,* ed. B. Grzimek, Band 13, pp. 342-367. Zürich: Kindler. English ed. 1972. Duikers, dwarf antelopes, and tragelaphines. *In: Grzimek's animal life encyclopedia,* ed. B. Grzimek, vol. 13, pp. 308-330. New York: Van Nostrand Reinhold Co.

Walther, F.R. 1969. Flight behaviour and avoidance of predators in Thomson's gazelle (*Gazella thomsoni* Günther, 1884). *Behaviour* 34:184-221.

Walther, F.R. 1972a. Social grouping in Grant's gazelle (*Gazella granti* Brooke, 1872) in the Serengeti National Park. *Z. Tierpsychol.* 31:348-403.

Walther, F.R. 1972b. Territorial behaviour in certain horned ungulates, with special reference to the examples of Thomson's and Grant's gazelles. *Zool. Afric.* 7:303-307.

Walther, F.R. 1973a. Round-the-clock activity of Thomson's gazelle (*Gazella thomsoni* Günther, 1884) in the Serengeti National Park. *Z. Tierpsychol.* 32:75-105.

Walther, F.R. 1973b. On age class recognition and individual identification of Thomson's gazelle in the field. *J. Sth. Afr. Wildl. Mgmt. Ass.* 2:9-15.

Walther, F.R. 1974. Some reflections on expressive behaviour in combats and courtship of certain horned ungulates. *In: The behaviour of ungulates and its relation to management,* ed. V. Geist and F.R. Walther, pp. 56-106. IUCN Publ. No. 24. Morges: IUCN.

Walther, F.R. 1977a. Sex and activity dependency of distances between Thomson's gazelles (*Gazella thomsoni* Günther 1884). *Anim. Behav.* 25:713-719.

Walther, 1977b. Artiodactyla. *In: How animals communicate,* ed. T.A. Sebeok, pp. 655-714. Bloomington: Indiana Univ. Press.

Walther, F.R. 1978a. Forms of aggression in Thomson's gazelle: their situational motivation and their relative frequency in different sex, age, and social classes. *Z. Tierpsychol.* 47:113-172.

Walther, F.R. 1978b. Quantitative and functional variations of certain behaviour patterns in male Thomson's gazelle of different social status. *Behaviour* 65:212-240.

Walther, F.R. 1978c. Mapping the structure and the marking system of a territory of the Thomson's gazelle. *E. Afr. Wildl. J.* 16:167-176.
Walther, F.R. 1979. Das Verhalten der Hornträger (Bovidae). *Handb. Zool.* 8 10(30):1-184.
Walther, F.R. 1981. Remarks on behaviour of springbok, *Antidorcas marsupialis* Zimmermann 1790. *D. Zoolog. Garten* 51:81-103.
Weber, M. 1928. *Die Säugetiere.* Jena: Fischer.
Zohary, M. 1962. *Plant life of Palestine.* New York: Ronald Press.

Index

Other Noyes Publications

IGUANAS OF THE WORLD
Their Behavior, Ecology, and Conservation

Edited by

Gordon M. Burghardt
University of Tennessee
Knoxville, Tennessee

A. Stanley Rand
Smithsonian Tropical Research Institute
Balboa, Panama

Animal Behavior, Ecology, Conservation and Management Series

This book is the first to give an in-depth interdisciplinary treatment of the ethology, ecology, and conservation of any reptile group. Iguanas are large herbivorous lizards economically important as food and for recreational hunting. Scientifically, they are pivotal in understanding the evolution of land vertebrates and the transitions needed for mammal and bird radiation from reptilian ancestors. Yet they are rapidly declining in numbers and several species have become extinct; several more are threatened or endangered. Only in recent years have efforts been made to study iguana behavior, reproduction, and habitat requirements. This volume brings together studies on species throughout their range–the West Indies, the Galapagos, Fiji, Central and South America.

This volume is divided into sections dealing with the diversity and distribution of iguanas, feeding and food utilization, reproduction, social systems, communication and conservation. A distributional and taxonomic map is included. **Color plates of all the genera are also included.**

ISBN 0-8155-0917-0 (1982) **480 pages**

Other Noyes Publications

WOLVES OF THE WORLD

Perspectives of Behavior, Ecology, and Conservation

Edited by

Fred H. Harrington
Mount Saint Vincent University
Halifax, Nova Scotia
Canada

Paul C. Paquet
Portland State University
Portland, Oregon

This book brings together the latest worldwide status of the behavior, ecology, and conservation of wolves by authorities from around the world. The North American section presents the latest information available on wolves from Alaska through Canada and Minnesota. The section on the wolves in Eurasia presents a description of these animals poised near the brink of extinction, and covers the wolves in the USSR, Northern Europe, Sweden, Italy, Iran and Israel. Captive wolf behavior and sociology is also discussed separately. The section on conservation examines in detail what must be done to maintain the current range of wolves worldwide. The book provides a telling contrast between wolves as they were in the past throughout the world, and as they may become throughout the world in the future.

It is a most valuable reference work that should be on the book shelf of all those interested in, and studying, animal behavior, sociobiology, wildlife management, ecology, conservation, zoology and animal science.

ISBN 0-8155-0905-7 (1982) **474 pages**